HIGH ENERGY
GAMMA-RAY EXPERIMENTS

Proceedings of the Third Workshop on

Science with the New Generation of

HIGH ENERGY GAMMA-RAY EXPERIMENTS

Cividale del Friuli, Italy 30 May – 1 June 2005

Editors

Alessandro De Angelis

INFN, University of Udine and IST Lisboa

Oriana Mansutti

INFN and University of Udine

World Scientific

NEW JERSEY • LONDON • SINGAPORE • BEIJING • SHANGHAI • HONG KONG • TAIPEI • CHENNAI

Published by

World Scientific Publishing Co. Pte. Ltd.
5 Toh Tuck Link, Singapore 596224
USA office: 27 Warren Street, Suite 401-402, Hackensack, NJ 07601
UK office: 57 Shelton Street, Covent Garden, London WC2H 9HE

British Library Cataloguing-in-Publication Data
A catalogue record for this book is available from the British Library.

Proceedings of the Third Workshop on Science with the New Generation of
HIGH ENERGY GAMMA-RAY EXPERIMENTS

ISBN 981-256-813-1

Printed in Singapore by B & JO Enterprise

CONTENTS

INTRODUCTION

This is the third workshop of a series on High Energy Gamma-ray Experiments, following the conferences held in Perugia 2003 and Bari 2004. The previous workshops were attended by more than 50 scientists in the field. The aim is to update the current and planned research for the space-borne and ground-based experiments dedicated to the observation of the Gamma-Ray Sky. Among the participants are both veterans of the first dedicated gamma-ray missions (like SAS-2 COS-B), and young students entering the fascinating field of gamma-ray astrophysics, captivated in the '90s by the results and the revolution performed by the Compton Gamma-Ray Observatory, CGRO (1991–2000).

Because of the Earth's atmosphere shield, gamma-ray astronomy could not develop until it was possible to get detectors on spacecrafts, and to build large and sensitive ground-based detectors. The first gamma-ray telescope carried into orbit, on the Explorer XI satellite in 1961, picked up less than 100 cosmic gamma photons. Additional gamma-ray experiments flew on the OGO, OSO, Vela, and Cosmos series of satellites, and in the 80's we had the first missions completely dedicated to the exploration of the gamma universe. New unexpected phenomena were discovered by satellite observations, as for example gamma-ray bursts and gamma-ray emission from blazars.

The expectations from the next generation of ground-based and space-borne high energy gamma-ray projects are very exciting both for quantity and quality of data that will be produced. The great vitality of this field in the current "post-EGRET era" is witnessed by the large number of on-going and forthcoming gamma-ray experiments, complementary to various cosmic- ray, neutrino and astroparticle projects, and to X-ray missions. The scientific community interested in gamma-ray astrophysics and astroparticle physics is rapidly growing, and this workshop series aims at exchanging periodically ideas and results, and optimizing the analysis of data.

The workshop is organized by the Physics Department of the Udine University and by I.N.F.N.

Part I

Detectors for High-Energy Gamma-Rays

FIRST RESULTS FROM THE MAGIC EXPERIMENT

DENIS BASTIERI FOR THE MAGIC COLLABORATION*

Dipartimento di Fisica, Università di Padova
& INFN Sezione di Padova,
Via Marzolo 8 – 35131 Padova, Italy
E-mail: bastieri@pd.infn.it

The MAGIC Telescope after a full year of commissioning phase is now successfully taking data since October 2004. Many technical innovations contribute in lowering its energy threshold well below 100GeV. Among them, the all-aluminium parabolic reflecting surface that, with its 236 m^2, makes it nowadays the biggest Cherenkov telescope and a smart trigger able to cope with the increased background rate. In this contribution, along with a brief overview of the scientific goals of MAGIC, details will be given on the construction and performance of the telescope, highlighting its first observational results and the potential for new physics.

1. Introduction

The MAGIC Telescope† was built at 2,200 m a.s.l. at the Observatorio del Roque de Los Muchachos (ORM) (28.8° N, 18.8° W), the site that, together with the Observatorio del Teide, makes up the so-called European Northern Observatory (ENO), a renown site for the high astronomical quality of the sky, due to the very low humidity and to minimal air turbulences.

In the construction of MAGIC, the **M**ajor **A**tmospheric **G**amma **I**maging **C**herenkov Telescope, the collaboration aimed at lowering as much as possible the current energy threshold of ground detectors, in order to open for observation a new energy window of the electromagnetic spectrum between 30 and 250 GeV.

This energy window has as lower limit the upper energies observable by satellite-borne detectors. These experiments have to clash with the poor statistics of high-energy events and with the limited amount of weight available to calorimetric identification. This limit, of $\sim$ 10 GeV, was obtained

*`http://magic.mppmu.mpg.de/collaboration/members/index.html`
†`http://magic.mppmu.mpg.de/`

on-board CGRO with EGRET and will be pushed somewhat up with the oncoming launches of AGILE[1] and GLAST[2].

Instead, ground-based experiments, and especially IACT (**I**maging **A**tmospheric **C**herenkov **T**elescopes) like MAGIC, are pulling the upper limit down. Ground experiments detect the shower development of primary particles in our atmosphere and not the primary particles themselves, as satellite-borne experiments do, thus, they have to rely on Montecarlo simulations in order to calibrate the algorithms for the energy reconstruction and particle identification. On the other hand, they are not limited by statistics, as their effective area is not the actual detector size, as for satellite ($\sim 1 \div 10\,\mathrm{m}^2$), but the cross section of the developing shower ($10^{4\div 5}\,\mathrm{m}^2$).

It can be shown that the energy threshold of IACTs roughly scales down with the reflecting surface and correlated quantities (like reflectivity, conversion efficiency, ...). For this reason, modern IACT projects are all endowed of enormous surfaces: HESS‡, CANGAROO III§ and VERITAS¶ feature $10 \div 12\,\mathrm{m}\,\varnothing$ dishes. Besides them MAGIC, with a $17\,\mathrm{m}\,\varnothing$ dish and a total reflecting surface of $236\,\mathrm{m}^2$. Nevertheless, having huge surfaces is not enough to lower energy thresholds: as the *solar farm* case showed, one has also to cope with background, increasing consequently with the bigger area. In this contest, having a smart trigger that can reject a background fraction even at hardware level can be essential.

2. Scientific Subjects

Taking the burden of lowering the energy threshold has, as its counterpart, a new realm of observations[3]. The most prominent case are AGNs, that at EGRET energies are counted by dozens and populate different redshifts, while for older IACT detectors (like Whipple and HEGRA) did not show up beyond $z \approx 0.03$. These differences can arise from an internal cut-off of the AGN power-law spectrum between 20 and 250 GeV, or from interaction of the emitted gammas with the extragalactic background light. In this context MAGIC, that can detect in 50 hours at a 5σ-level a source with a *Crab-like spectrum* and an integral flux of $10^{-10}\,\gamma\,\mathrm{cm}^{-2}\mathrm{s}^{-1}$ as low as at 50 GeV, should observe plenty of AGNs and even discriminate between the two models (see Fig. 1 for an updated comparison of sensitivities of various experiments).

‡`http://www.mpi-hd.mpg.de/hfm/HESS/HESS.html`
§`http://icrhp9.icrr.u-tokyo.ac.jp`
¶`http://veritas.sao.arizona.edu`

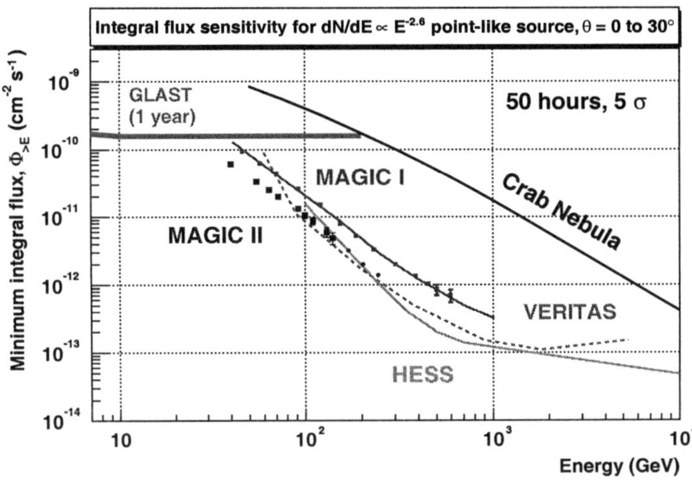

Figure 1. Predicted sensitivities for some operating and proposed detectors. Note the wide overlap between GLAST and present Cherenkov telescopes. As far as MAGIC is concerned, the solid, red line represents the predictions made by the full Montecarlo simulation, in good agreement with the sensitivity calculated from the first observations. Blue dots are the expected sensitivity for MAGIC II, a *clone* of the current MAGIC, that will be built at 85 m of distance from MAGIC. Start of operation for MAGIC II is envisaged for the beginning of 2007, just before the scheduled launch of GLAST.

Moreover, EGRET legacy consisted of plenty of unidentified sources, and observation stretching up the energy window could help knowing them. AGNs, alongside with other gamma emitters such as supernova remnants (SNRs) and microquasars can all fit in this class: a thorough study could bridge the energy gap between satellite and ground detectors and identify the main sources of cosmic rays up to an energy of 10^6 GeV.

Related to AGNs, or better to a catalogue of observed AGNs at different redshifts, is the measurement of many cosmological parameters. The *Hubble parameter* can have its more precise measurement in this way.

The interesting phenomenon of GRBs can be also a subject for MAGIC that, due mainly to its stiff and lightweight supporting cradle, can be re-pointed in less than 20 seconds. An early satellite follow-up, as the one provided by SWIFT, gives MAGIC the unique opportunity of an early follow-up ($\lesssim 30$ s) at energies well below 100 GeV.

Finally, a low energy threshold, virtually inside satellite energy window, can help in calibrating the energy reconstruction methods that now deeply rely on Montecarlo techniques. Satellite experiments and their direct detection of cosmic rays can thus provide a sort of *test beam* for MAGIC[4].

3. The MAGIC Telescope

The MAGIC telescope is an international collaboration that among its main components enrols Germany, Spain and Italy. MAGIC fits nicely in a frame of continuous monitoring in the hundred GeV region for both hemispheres, since the other three *second generation* collaborations are situated in Australia (CANGAROO), southern US (VERITAS) and Namibia (HESS). It features a reflecting surface of parabolic shape F/1 and $236\,\mathrm{m}^2$ of size, made up with 964 aluminium mirrors, of $80 \div 90\%$ reflectivity. The collected light is focused in a camera equipped with 576 6-dynode compact photomultipliers, featuring $\sim 20\%$ of quantum efficiency in the $300 \div 500\,\mathrm{nm}$ band[5].

A two-level trigger system[6] takes care of removing most of the background light, and start the conversion to digital data. Pixel signals are stretched and converted with a high and a low gain to enlarge the dynamic range. The final digitisation is done using 300 MHz 8-bit flash ADCs. The maximum readout rate is around 1kHz, for a total of 80GB of data collected on the longest nights ($\sim$ 11 hours). The typical readout rate of collected data is instead much lower, $250 \div 350$ Hz, taking into account also a 50 Hz rate of calibration events.

The telescope features also a very lightweight mount, mainly made with a *filigree* structure of carbon fibre tubes jointed together onto aluminium ball-nodes. The resulting structure is very stiff, reacting to 10 m/s wind gushes with a $\sim$ 3 cm maximal displacement of the camera. On the other hand, being so lightweight, MAGIC can promptly react to *ToO* (**T**arget **o**f **O**pportunity) such as GRB alerts, as was the case of GRB 050723, simultaneously observed with SWIFT during the flaring activity.

4. First observations of sources and performance on data

The first sources observed were the Crab Nebula and MKN 421. Both of these sources are a sort of *standard* sources: the Crab is a *plerion*, a SNR hosting a pulsar in its core, and has a steady flux that can be used to calibrate detectors. MKN 421 is a very close AGN ($z = 0.03$) and, while not being steady, stays from time to time in a high state with fluxes few times bigger than the Crab one. They were observed even in the commissioning phase in Winter 2003/2004 (the Crab Nebula) and during Spring 2004 (MKN 421) and both revealed signals well above the 5σ level.

The Crab and MKN 421 were also observed later on, during the first semester of data acquisition. In addition, another AGN, 1ES1959+650, farther than MKN 421, was detected even if it was quiescent.

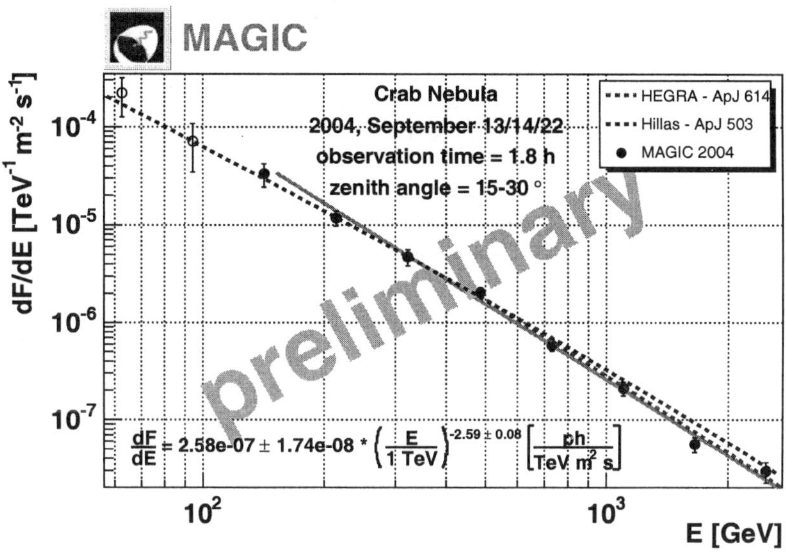

Figure 2. Preliminary Crab spectrum obtained after analysing 2 hours of data taken September between 15° and 30° of zenith angle. The grey line corresponds to the fitted formula.

4.1. *The Crab Nebula*

The *Crab Nebula* is nowadays the standard candle for VHE (Very High Energy) gamma astronomy, having a steady flux since many years. In this energy range the emission is mainly due to the Inverse Compton Scattering of the synchrotron radiation. The observation at low energies should allow the study of the Compton peak, of interest for many theoretical models.

Given the importance of the Crab Nebula, mainly for calibration purpose and for the aforementioned Compton peak, MAGIC enlisted the Crab as its first scheduled object, even during the first semester of the physics campaign.

Collected and analysed data showed the presence of a signal even at low energies (< 100 GeV) (see Fig. 2). The entity of systematic errors is still under evaluation, even if we do not expect a substantial modification of the proposed spectrum especially at high energies.

4.2. *Markarian 421*

While on commissioning phase, on April 22nd, 2004, MKN 421 was in a high state. Data collected during that night, with MAGIC pointing at

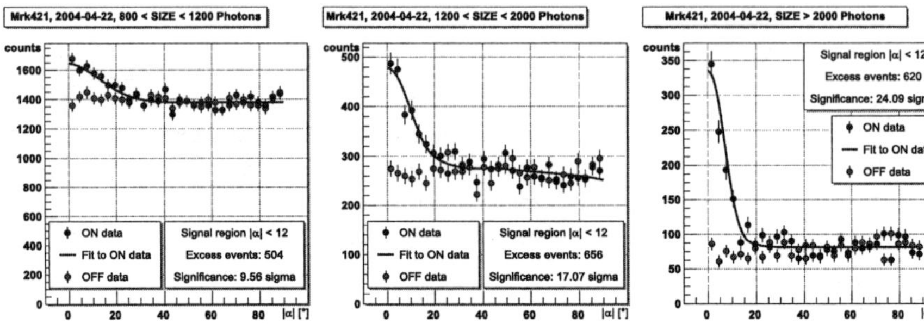

Figure 3. Data from 155-minutes of acquisition taken on April 22nd, 2004 with MAGIC pointing at Mkn 421. The histogram of *alpha* for on- and off-source observations, are relative to different bin in *size*: 800–1200 (left), 1200–2000 (middle) and more than 2000 photons (right).

the source, are shown in Fig. 3. MAGIC, as all IACTs, records for each event the directions of the Cherenkov photons emitted during the shower development. The image is then analysed using a well-established technique that exploits the so-called *Hillas parameters*[7].

Two Hillas parameters have a particular importance in the figure: *alpha* and *size*. *Alpha* is related to the actual direction of the primary particle that initiated the shower, thus an excess in *alpha* must be seen in the direction of a source. *Size* is the number of photons making up the image and is related to the energy of the primary particles (the more energetic is the particle, more photons appear in the image).

The three figures show that the Mkn 421 flare was well detected by MAGIC at different energies, and the excess seen in the first figure with *size* ranging from 800 to 1200 photons is consistent with an energy well below 100 GeV.

Observation made afterward, during the physics campaigns, featured a joint observation with HESS. On December 2004, the X-ray satellite RXTE reported the high state of Mkn 421. The AGN was repeatedly observed by many ground detectors in many different wavelengths. Among them MAGIC and HESS that were able, on December 18th, 2004, to observe simultaneously the source, allowing:

- a mutual cross calibration of the two experiments, that could help in understanding and reduce systematic errors;
- a simultaneous observation in two different energy ranges, as the source was visible to the two experiments at different zenith angles.

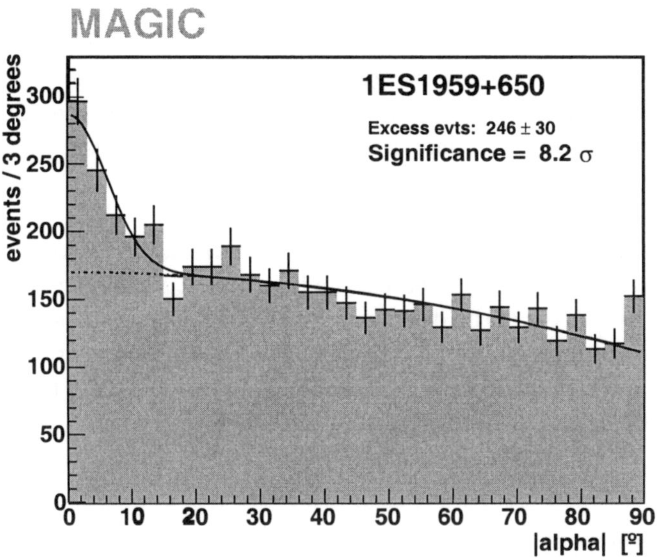

Figure 4. *Alpha* plot of 1ES1959+650 after cuts on image parameters. The graph shows the second-order curve used for estimating the background at low *alpha* (up to 9°).

4.3. *1ES1959+650*

The AGN 1ES1959+650 is an X-ray peaking *BL Lacertae* object selected from the Einstein Medium Sensitivity Survey. It is hosted by an elliptical galaxy at a redshift of $z = 0.047$.

MAGIC has observed VHE gamma-ray emission from 1ES1959+650 during six hours in September and October 2004. The observations were carried out alternated with the *Crab Nebula*, whose data were used as reference source for optimising gamma/hadron separation and for flux comparison. The data analysis shows VHE gamma-ray emission of 1ES1959+650 with $\sim 8\sigma$ significance, at a time when the AGN was reported to be quiescent in both optical and X-ray wavelengths (see Fig. 4).

An integral flux above $\sim$ 180 GeV of about 20% of the Crab was obtained. The light curve, sampled over 7 days, shows no significant variations. The differential energy spectrum between 180 GeV and 2 TeV can be fitted with a power law of index -2.72 ± 0.14. The spectrum is consistent with the slightly steeper spectrum seen by HEGRA at higher energies, also during periods of low X-ray activity.

5. Conclusions

The MAGIC telescope has started its regular operation. Designed to be a pioneering telescope for low-energy gamma astronomy, it is now giving its first results. Several sources were detected, among them the *Crab Nebula* that presumably can be studied even close to the Compton peak and the *medium*-far AGN 1ES1959+650, that even if quiescent and at a flux level of $\sim \frac{1}{10}$ Crab was well detected in 6 hours.

New analysis tools, including timing info, are under study, to work at lower energies and further enhance the instrument sensitivity.

Given the success of the detector, the MAGIC Collaboration started the construction of a second detector (MAGIC II) to be built at a distance of 85 m from the current MAGIC telescope. Start of operation for MAGIC II is envisaged for the beginning of 2007, just before the scheduled launch of GLAST.

Acknowledgments

The construction of the MAGIC Telescope was made possible by the financial contributions of the Italian INFN, the German BMBF and the Spanish CICYT, to whom goes our grateful acknowledgement. We would also like to thank the IAC for the excellent working conditions provided at El Roque de los Muchachos.

References

1. M. Tavani et al. (2001), Procs. Symp. *Gamma 2001*, S. Ritz, N. Gehrels, C.R. Shrader eds., AIP Conf. Proc. 587, 729.
2. P. Michelson et al. (1999), Response to AO 99-OSS-03, GLAST Large Area Telescope, Flight Investigation: An Astro-Particle Physics Partnership Exploring the High-Energy Universe Volume 1: Scientific and Technical Plan.
3. A. Saggion and D. Bastieri, *"The Observational Energy Gap between 10 and 300 GeV"*, *Mem. Soc. Astro. It.* 73/4 (2002), pp. 812–821.
4. D. Bastieri *et al*, *Astropart. Phys.* 23/6 (2005), pp. 572–576, arXiv:astro-ph/0504301.
5. A. Ostankov *et al*, *Nucl. Inst. & Meth.* A 471 (2001), pp. 188–191.
6. D. Bastieri *et al*, *Nucl. Inst. & Meth.* A 461 (2001), pp. 521–523.
7. A.M. Hillas (1985), Procs. 19th ICRC, La Jolla, USA 1985, vol. 3, p. 445–448.

H.E.S.S.

P. VINCENT FOR THE H.E.S.S. COLLABORATION
Laboratoire de Physique Nucléaire et de Hautes Energies, IN2P3/CNRS, Universités Paris VI & VII, 4 Place Jussieu, F-75252 Paris Cedex 05, France

The first phase of the H.E.S.S. experiment is now fully operational since the end of 2003 but observation has started in mid 2002. After a systematic study of know TeV sources in the southern hemisphere, the H.E.S.S. collaboration has undertaken a scan of the inner region of the galactic plane. It also accomplished observations of many other galactic and extragalactic objects. We report here on the observation periods between 2003 and 2004.

1. Introduction

The H.E.S.S. experiment[1] is an array of four Imaging Atmospheric Cherenkov Telescope installed in the Khomas district in Namibia. Each telescope[2] has a large Davis-Cotton mirror of 108 m^2 area fixed on a alt-azimuth mount. The focal plane is equipped with a camera[3] of 960 photo-multipliers for a total field of view (FoV) of 5°. This large FoV is specially propitious to the observation of extended sources like supernovae remnant (SNR) and also favourable to observations in scan mode or to search form new sources of high energy gamma rays. The angular resolution of individual shower reconstruction is about 0.1° and the precision in the determination of the source position is better than 1'. With a threshold of about 100 GeV and a sensitivity at the level of a percent of a Crab-like source; this experiment has reached its objectives concerning performance. On the site of operation, an online analysis based on a preliminary online calibration chain gives an indication on the significance of the tracked objects after one hour. However, all the analyses presented here or published in reviews are based on two independent offline calibration chains developed in parallel to check our understanding of the apparatus and also use three different methods of analysis. The first one is based on Hillas parameterisation of shower images followed by a geometrical reconstruction of the shower axis[4]. The two other methods[5,6] used full two or three dimensional models to predict the Cherenkov light emitted by gamma-ray initiated air showers and perform a fit to each experimental image to find the best parameters of the initial photon (see also [7] for comparison of these three methods). The efficiency of the source reconstruction is better than 50% in the inner 3° region of the FoV and some analyses based on the analytical model have demonstrated the possibility to reconstruct sources located at the edge of the total FoV. Between 2003 and 2004, the H.E.S.S. collaboration have tracked more than one hundred targets for a total of about 1,000 hours of observation time per year. A scan of a small inner region of our galaxy has been carried out[8] in the range of ±30° of galactic longitude and ±3° in latitude. The result of this scan is shown in figure 1. A total of eleven sources have a significance higher than 6 σ (white labels), and also seven hot spots at more than 4 σ are present in this scan (in yellow). Galactic sources, outside this region, were also observed

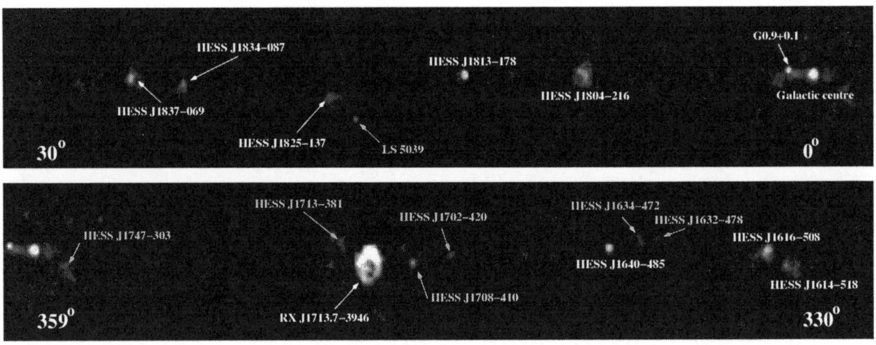

Figure 1. H.E.S.S. scan of the central region of the Milky Way. The top figure shows the scan from −30° to +2° and bottom scan from −2° to +30°.

and detected; whereas some source remains undetectable like, for instance, the SNR **SN1006** and the pulsar wind nebula (PWN) **PSR B1706-44**. They has been observed and upper limits have been published[9,10]. Extragalactic observations has been carried out for several blazars as well as non-blazar targets bringing new results. These observations have been used to constrain the level of extragalactic background light[11]. In this paper we will not cover all these topics; the reader is invited to refer to published papers in references and also to the proceeding of the ICRC conference[12] for the latest H.E.S.S. result.

2. Galactic sources

The H.E.S.S. observations have allowed for the first time in this domain the source morphology. This is particularly important to understand the location of the gamma production and to identify the surrounding environment of these sources. To carried out such analysis one must achieve a high position resolution as well as collecting a sufficient data sample. With the accuracy of the position reconstruction and the sensitivity of the H.E.S.S. apparatus these two conditions are fulfilled. Figure 2 show two spectac-

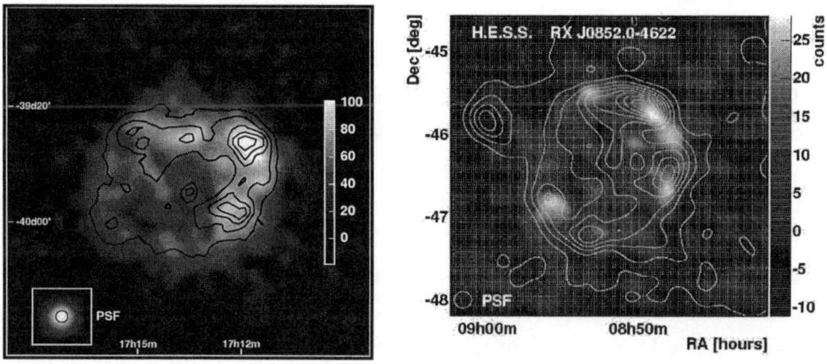

Figure 2. H.E.S.S. excess map of RX J1713-3946 (left) and RX J0852.0-4622 (right). The superimposed white lines indicate the contour in X-rays from ASCA (RX J1713) and ROSAT (Vela). The image is smoothed with a Gaussian of 2 arcmin standard deviation, the linear colour scale is in units of excess counts.

ular demonstrations of this performance. On left, a $2^\circ \times 2^\circ$ field of view centred on the SNR **RXJ 1713-3946**[13] is shown. Initially discovered by the satellite ROSAT in an all-sky-survey X-rays at about 1 kpc, this object has been later claimed as source of TeV gamma rays by the CANGAROO collaboration[14]. The distributed gamma emission is similar to a shell morphological structure with no clear evidence of any emission from the central region. It demonstrate that shocks generated by supernovae can accelerate charged particles up to multi-TeV energies producing secondary high energy gammas by interaction with the interstellar medium in case of protons or inverse compton process for electrons. The overall shell structure coincides with that seen in X-ray contours (from ASCA) which indicates that the two type of emission have a similar origin. The overall gamma-ray flux corresponds to about 66% of the flux of the Crab Nebula as measured with H.E.S.S. This distribution shows a significant asymmetry between a brightest part (with an excess of 20σ) in the West and northwestern region of the map and the southeastern region. It is believed that in the highest emitting region, interaction with dense molecular clouds enhance the gamma ray production. **RX J0852.0-4622** (Vela Junior), previously announced by the CANGAROO collaboration at TeV energies[15], is an other interesting example of shell structure emitting object. The right of figure 2 shows the excess map reconstructed by the H.E.S.S. detector[16]. From the galactic scan a total of eighteen sources, sixteen of which are news, have been discovered. Three of them were clearly identified in this scan and associated with known SNR in other wavelength. **HESS J1640-465** is a marginally extended source with respect to the H.E.S.S. point spread function. Its flux is close to 9% of the Crab above 200 GeV, with a spectral index of 2.42 ± 0.15. Its position is closely correlated with the SNR G338.3-0.0. It is also compatible with 95% confidence contour of the EGRET source 3EG J1639-4702, and its spectrum extrapolated down to 1 GeV is compatible. Two other SNR candidates are **HESS J1804-216**, a bright source at 20% of the Crab and **HESS J1834-087**, a faint gamma emitter at 5% of the Crab. They are both extended with a size of 22 and 12 arc minutes respectively with similar spectral index. The first can be associated with the South-Western part of the shell of G008.7-00.1, but it also coincides with the pulsar PSR J1803-2137. The second positionally coincides with the shell-type SNR G023.3-00.3.

Stelar object were gamma rays are generated in pulsar wind nebulae, similar to the Crab Nebula, are another class of TeV emetters. Among them, the source termed **HESS J1825-137**, initially announced as an unidentified TeV source, has been recently associated to the close but not clearly coincident pulsar PSR B1823-13[17]. This pulsar, with an estimated distance of about 4 kpc, reveals an asymmetric spatial distribution of TeV gamma rays, peaking at the pulsar position and decaying toward the South. This behaviour is remarkably similar to the situation in X-rays, where the pulsar is surrounded by a nebula with a pronounced tail toward the South, except that the scales are different. Relativistic electrons and positrons are presumed to be ejected symmetrically from the pulsar, but the north jet was crushed by a reverse shock rebounding when the supernova shock wave had piled up enough interstellar material. When expanding into an inhomogeneous medium, the reverse shock from the (lower-density) South side may not yet have reached the PWN[18]. The SNR **MSH 15-52** was first discovered in radio observations[19] . It reveals a roughly circular supernova shell with a bright emission in the North-West rim and a fainter one in the South-East. At the centre of the shell, a 150 ms X-ray pulsar (PSR B1509-58) remains from the supernova explosion. The pulsar is surrounded by a diffuse nebula of unpulsed X-ray emission, similar to that observed in the Crab nebula. The Chandra X-ray image[20] shows a jet-like structure emerging from the pulsar with a collimated outflow arc-like structures in the North of the pulsar. In 2004, about 22 hours of data were accumulated on

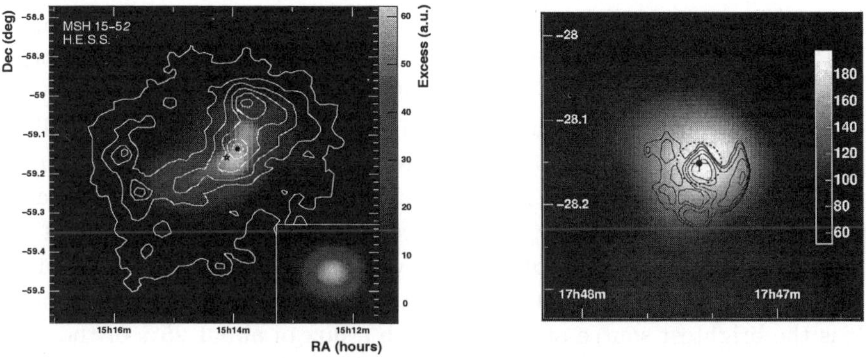

Figure 3. H.E.S.S. excess map of MSH 15-52 (left) and G00.9+00.1 (right). The superimposed white lines indicate the contours in X-rays from ROSAT. The image is smoothed with a Gaussian of 2 arcmin standard deviation, the linear colour scale is in units of excess counts.

MSH 15-52, showing a clear TeV gamma-ray signal[21] with a significance of 25 sigma and a flux of roughly 15% of that observed for the Crab Nebula. The TeV gamma rays show a clearly resolved structure (Fig. 3, left), with an elongated and slightly curved emission region, similar to the morphology of the X-ray jet. In the North-West direction, the source has a characteristic width of 6' (assuming a Gaussian source profile), and in the orthogonal direction a width of about 2', The centre of gravity of the gamma-ray emission is displaced from the pulsar. In 2003, first observations of the Galactic Centre region with two H.E.S.S. telescopes already showed a faint signal at the location of **G00.9+00.1**[22]. In 60 h of follow-up observations between March and September 2004, with all four H.E.S.S. telescopes, a highly significant signal is detected (13 sigma), consistent with the position of the core of the SNR (Fig. 3, right). Within the angular resolution of H.E.S.S., the source is point-like, with an emission region smaller than 1.3' (rms). The location of the source and the size limit strongly suggest the PWN as the source of the VHE gamma-ray signal, rather than the supernova shock wave.

In the vicinity of the binary system PSR B1259-63/SS 2883, discovered by the H.E.S.S. collaboration[23], the experiment, equipped with camera of 5° of FoV has, in 2004, detected simultaneously a second object which is unidentified in any other wavelength. This new object, called **HESS J1303-631**, exhibits an hard spectrum of 2.4 ± 0.2 and a strong signal with a flux of 17% of the Crab. It has been observed in 2004 with a significance of 13 σ level at 0.7 degrees to the North of the binary system[24]. Several pulsars are known with positions close to HESS J1303-631. However, from the energetics point of view, most of them are rouled out appart PSR J1301-6305 which remain a interesting candidate. Up to now, no other plausible candidates such as dense gas clouds illuminated by cosmic rays or dark matter annihilation clumps match this signal. More sensitive follow-up measurements in other wavelength regimes might help to pin down the physics of this new source. Another unidentified source from the galactic scan is **HESS J1614-518**, with a reconstructed mean size of 12±1 arc minutes with elliptical morphology. It was identified at galactic longitude 331.52 ± 0.03 and latitude -0.58±0.02, close to the pulsar PSR J1617-5055. It is the brightest source of the scan with a flux of about 25% of the Crab Nebula flux. Its spectrum fit a power law with photon index 2.46 ± 0.20. The observation of a region of 10 arc minutes around the location of this source by the Chandra X-ray satellite does not show any evidence of X-ray emission in the direction of this source. The galactic centre has been

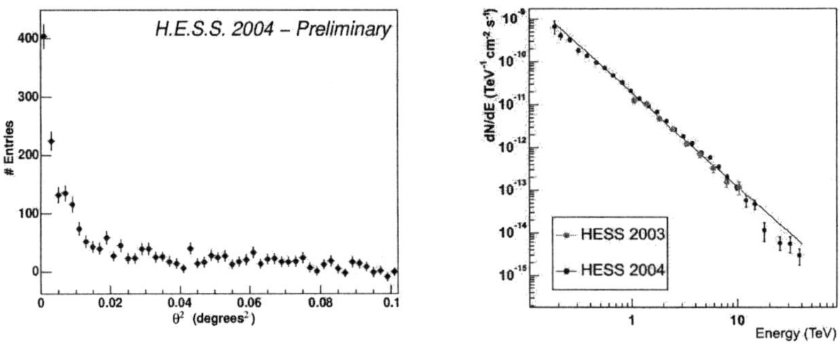

Figure 4. Excess of gamma-ray candidates in the H.E.S.S. data (left) in the direction of the Galactic centre and reconstructed energy spectrum (right).

observed continuously in 2003 and 2004[25]. The reconstructed spectrum exhibit the same spectral index and the flux seems stable over all time-scales. In the distribution of the gamma-ray excess we can see two components, a narrow profile close to the position of the source super-imposed on an extended tail. Various attemps to fit this distribution with predicted profile from dark matter annihilation failed, and more classic gamma-ray production processes must be envisaged to explain the observed signal.

3. Extra-galactic sources

From the H.E.S.S. site, **Mkn 421** ($z = 0.031$) is observed[26] at large zenith angle. With a reconstructed flux of about 3 Crab above 2 TeV, the data corresponding to 15 hours of observation allows a clear detection of 100σ. The light-curve in April and May exhibit significant intra-night variability. These observations were taken with a zenith angle between 60° and 67° which results a large collection area (2 km^2) at 10 TeV specially appropriate to study the effect of gamma ray absorption by extra-galactic background light on the end-point energy spectrum. The reconstruction spectral curvature can be described by a power law (2.1±0.1±0.3) with an exponential cut-off at 3.1(+0.5-0.4)±0.9 TeV or super-exponential cut-off at (6.25±0.4±0.9) TeV. The blazar **PKS 2005-489** ($z = 0.071$) has been detected[27] at a signal level of 6.7 σ in a sample of 24.2 hours live-time, which corresponds to 2.5% Crab flux. It is the first blazar independently discovered by H.E.S.S. at TeV energies and the second in the southern hemisphere. No evidence for variability at the scale of a year was seen. The energy spectrum ($\Gamma = 4.0 \pm 0.4$) corresponds to a particularly soft emission for a BL Lac. The detection of **PKS 2155-304**[28] ($z = 0.117$) was confirmed at more than 90 σ in 2004, with a soft power law shape. Two multiwavelength (MWL) campaigns took place in 2003 and 2004. In 2004 the variability was sufficient to suggest a correlation between X-rays and γ-rays activity and also to probe intra-night variability. MWL observations also indicate that PKS 2155-304 was in a low or quiescent state during some observations periods. Finally, the nearby radio Galaxy **M87** located at a red shift of 0.004 was detected at a significance level of 4.6σ[29]. The estimated flux of this source is between 0.5 and 1% of the Crab flux. With a jet oriented 20°-40° away from the observer's line of sight, this is a confirmation of the first non blazar extra-galactic source identified at TeV energies from the HEGRA collaboration[30], but more data are needed to state on a possible variability of the emission. Finally, the Seyfert galaxy **NGC 253** was observed during 28 hours and an upper limit has been recently published[31] well below the previous detection claimed by CANGAROO. Also, 19 BL Lacs and other AGNs at a read-shift between 0.002 and 0.3 has been observed and upper limits, between 1 and 3% of Crab flux, have been published[32] based on the analysis of 1 to 5 hours of observation time.

4. Conclusions

The H.E.S.S. experiment is the most sensitive ground based detector of high energy γ-rays above 100 GeV. Studies of the Southern sky at TeV energies have been underway for a few years now and reveal for the first time the morphology of TeV galactic sources like shell-type SNR, pulsar wind nebulae or the galactic centre. The scan of the galactic plane brought an abundance of new sources of which some without clearly identified counterparts. On the extra-galactic side, a non-blazar TeV source has been confirmed and multi-wavelength campaign on AGN are undertaken with H.E.S.S. All these results stimulate the research and understanding of VHE γ-rays sources.

Acknowledgements

We would like to thank all the institutions that support the H.E.S.S. project and the PHD students, the PostDoc and the technical staff who contribute in a major part to the success of this experiment.

References

1. W. Hofmann et al., 2003, Proc. 28th ICRC, Tsukuba, p. 2811.
2. K. Bernlöhr et al., Astropart. Phys. 20 (2003) 111.
3. P. Vincent, et al., 2003, Proc. 28th ICRC, Tsukuba, p. 2887.
4. W. Benbow, 2004, 3rd Gamma-ray Inter. Symp. (Heidelberg), 611.
5. M. de Naurois et al., 2003, Proc. 28th ICRC, Tsukuba, Vol. 5, p. 2907.
6. M. Lemoine-Goumard et al., 2004, 3rd Gamma-ray Inter. Symp. (Heidelberg), 697.
7. M. de Naurois, Proc. of Cherenkov 2005, Palaiseau, (to be published).
8. F. Aharonian, et al. (H.E.S.S. coll.), Science 307 (2005) 1938.
9. F. Aharonian, et al. (H.E.S.S. coll.), Astron. Astrophys. 437 (2005) 135.
10. F. Aharonian, et al. (H.E.S.S. coll.), Astron. Astrophys. 432 (2005) 9.
11. F. Aharonian, et al. (H.E.S.S. coll.), 2005, astro-ph/0508073.
12. W. Hofmann et al., 2005, Proc. 29th ICRC, Pune, India, (to be published).
13. F. Aharonian, et al. (H.E.S.S. coll.), Nature 432 (2004) 75.
14. Enomoto, R., et al., Nature, 416, 823, 2002.
15. Watanabe S., et al. 2003, Proc. 28th ICRC, Tsukuba, p. 2397.
16. F. Aharonian, et al. (H.E.S.S. coll.), Astron. Astrophys. 437 (2005) 7.
17. F. Aharonian, et al. (H.E.S.S. coll.), Astron. Astrophy, (to be published).
18. J. M. Blondin et al., Astrophys. Journal, Vol 563, Issue 2, p. 806.
19. J. L. Caswell et al., Royal Astron. Soc., vol. 195, Apr. 1981, p. 89.
20. B. M. Gaensler et al., Astrophys. Journal, Vol 569, Issue 2, p. 878.
21. F. Aharonian, et al. (H.E.S.S. coll.), Astron. Astrophys. 435 (2005) 17.
22. F. Aharonian, et al. (H.E.S.S. coll.), Astron. Astrophys. 432 (2005) 25.
23. F. Aharonian, et al. (H.E.S.S. coll.), Astron. Astrophys. 442 (2005) 4.

24. F. Aharonian, et al. (H.E.S.S. coll.), Astron. Astrophys. 439 (2005) 1013.
25. F. Aharonian, et al. (H.E.S.S. coll.), Astron. Astrophys. 425 (2004) 13.
26. F. Aharonian, et al. (H.E.S.S. coll.), Astron. Astrophys. 437 (2005) 95.
27. F. Aharonian, et al. (H.E.S.S. coll.), Astron. Astrophys. 436 (2005) 17.
28. F. Aharonian, et al. (H.E.S.S. coll.), Astron. Astrophys. 430 (2005) 865.
29. M. Beilicke et al., 2005, Proc. 29th ICRC, Pune, India, (to be published).
30. F. Aharonian, et al. (HEGRA coll.), Astron. Astrophys. 403 (2003) 1.
31. F. Aharonian, et al. (H.E.S.S. coll.), Astron. Astrophys. 442 (2005) 177.
32. F. Aharonian, et al. (H.E.S.S. coll.), Astron. Astrophys. 441 (2005) 465.

CANGAROO

MASAKI MORI FOR THE CANGAROO-II,III TEAM*
Institute for Cosmic Ray Research, University of Tokyo,
5-1-5 Kashiwanoha, Kashiwa, Chiba 277-0882, Japan
E-mail: morim@icrr.u-tokyo.ac.jp

The CANGAROO-III telescope system for very-high-energy gamma-ray astrophysics consists of four 10 m atmospheric Cherenkov telescopes located near Woomera, South Australia. The construction of the fourth telescope was completed in summer 2003, and stereoscopic observations with a local trigger system started in March 2004. Here we report on the status of the system and preliminary results from stereo observations.

1. Introduction

CANGAROO is an acronym for the Collaboration of Australia and Nippon (Japan) for a GAmma Ray Observatory in the Outback, and is a joint project of Japanese and Australian institutions. After the operation of the 3.8 m imaging Cherenkov telescope (CANGAROO-I) for 7 years, which was the first of this kind in the southern hemisphere, we constructed a new telescope of 7 m diameter (CANGAROO-II) in 1999 [1] next to the 3.8m telescope near Woomera, South Australia (136°47′E, 31°06′E, 160m a.s.l.). Then the construction of an array of four 10 m telescope (CANGAROO-III) was approved and as the first step the 7 m telescope was upgraded to 10m diameter in 2000, which is the first telescope of the CANGAROO-III array [2]. In the following years, we have constructed additional three 10m telescopes located at the corners of a diamond of 100m side with improved mirrors, cameras and electronics. After tuning, we have started observation with the full system in stereo mode in March 2004.

2. The CANGAROO-II telescope

After years of observations with the 3.8 m telescope, which was retrospectively called CANGAROO-I, a new budget to construct a whole new

*See `http://icrhp9.icrr.u-tokyo.ac.jp/` for the collaboration list.

telescope was approved in 1995 [1]. This telescope, CANGAROO-II, was equipped with a reflector [3] consisting of sixty spherical mirrors of 80 cm in diameter, which is approximately 7 m aperture, with a focal length of 8m. The base material of the mirror is CFRP (carbon-fiber reinforced plastic), which was newly developed for use in Cherenkov telescopes, and makes the reflector light and reduces gravitational deformation of the parabola shape. The altitude of each mirror is remotely adjusted by stepping motors. The 7 m telescope began operation in March 1999. In 1999 we obtained a new budget to construct an array of four 10 m telescopes, which is now called CANGAROO-III [2]. As the first step of CANGAROO-III, the 7m telescope was expanded to 10m by addition of 54 mirrors in March 2000, which is called 'T1' after installation of other telescopes. Basic specifications of T1 are given in Table 1.

3. Some results from CANGAROO-II

3.1. *Supernova remnant RX J1713.7-3946*

The first result from the CANGAROO-II telescope was published in Nature in April, 2002 [4]. A gamma-ray signal was found near the northwest rim of the supernova remnant (SNR) RX J1713.7-3946 (G347.3-0.5) using 2,332 min. on-source and 1,789 min. off-source data in 2000 and 2001. We argued that it could be evidence for the production of gamma-rays by non-electromagnetic process in the SNR using the observed spectrum which was expressed by a single power law and was not compatible with spectra assuming electromagnetic processes: inverse Compton scattering or bremsstrahlung. If it is interpreted as gamma-rays derived from pion decay process, this can be the first evidence of proton acceleration in supernova remnants, which might prove the SNR origin of cosmic rays. New H.E.S.S. results on this SNR [5] essentially support this interpretation.

Table 1. Basic specification of CANGAROO-III telescopes.

	T1	T2, T3, T4
Mount	Alt-azimuth	
Focal length	8m	
Number of mirrors	114 (57m^2 in total)	
Reflector type	Parabola	
Number of PMTs	552 (1/2")	427 (3/4")
Camera pixel size	$0.115°$	$0.168\ °$
Readout	TDC(CAMAC) & ADC	TDC(VME) & ADC
Point image size	$0.20°$ (FWHM)	$0.14 \sim 0.21°$ (FWHM)
Completion	2000.3	2002.3 (T2), 2002.11 (T3), 2003.7 (T4)

3.2. *The Galactic Center*

The Galactic center is one of the most attractive objects in the southern sky. Based on 4,000 min. (ON) and 3,400 min. (OFF) data with CANGAROO-II in 2001 and 2002, we detected a gamma-ray signal at energies greater than 250 GeV [6]. With detections by the Whipple [7] and H.E.S.S. [8] groups, now the Galactic center is an established TeV source. However, the CANGAROO-II spectrum seems to be steeper than that of H.E.S.S.: we will re-observe it with the CANGAROO-III stereo system in 2005.

3.3. *Supernova remnant RX J0852.0-4622*

This SNR, sometimes called as "Vela Jr." or G266.2-1.2, is a SNR located (in projection) at the southeast corner of the Vela SNR. It was discovered in 1998 using data from the ROSAT All-Sky Survey, and it is one of the few known SNRs which display strong non-thermal X-ray emission. With 5900 min. (ON) and 5300 min. (OFF) observations in 2001/2002 and 2003, we found a gamma-ray signal around the peak of the X-ray emission at energies greater than 500 GeV [9]. Again the spectrum seems to be steeper than that of H.E.S.S. [10], but the extended signal detected by the stereo observation in 2004 shows a harder spectrum when the whole emission region is taken into account (see 5.4).

4. The CANGAROO-III system

4.1. *Basic specification*

In 2002-2004 we have installed three new 10 m telescopes ('T2', 'T3' and 'T4') at the corners of a diamond of 100 m sides. This telescope array is called CANGAROO-III [2]. The major parameters of the CANGAROO-III telescopes are summarized in Table 1. The details of the system are described elsewhere [11]. Here we mention some recent updates [12].

4.2. *Star tracking accuracy*

Telescopes are driven by commands specifying elevation and azimuth angles sent to telescope controllers every 100 ms. Those values are computed from the celestial position of a target object by a control PC synchronized to a master PC equipped with a GPS receiver via `ntp` protocol.

Tracking accuracy of telescopes are monitored using CCD cameras mounted at the center of reflectors and faced to cameras. We can observe stars through the telescope focal ring when cameras are not installed and around the ring when cameras are installed with a wide camera lens. The root-mean-square deviation of displacement between observed and commanded positions is less than one arcminute, which is small enough compared with the size of a camera pixel (0.115° [T1] or 0.168° [T2, T3, T4]).

4.3. *Optical quality*

Each spherical mirror, made of laminates of fiber-reinforced-plastic and aluminium sheet [3], can have their direction adusted by two stepping motors. Mirrors of the first telescope were tuned using a distant light source [3]. For later (T2, T3, T4) telescopes we tuned mirrors to a common focus while stars were being tracked [13]. Focal images are captured by a CCD camera mounted at the center of the reflector before and after movements of motors and we can identify the displacement of each mirror by subtracting those images. By repeating this procedure on each mirror, we could tune all mirrors in a few nights per telescope. The point spread functions measured at the construction time are 0.20°, 0.21°, 0.14°, and 0.16° (FWHM). They are not as good as those of glass-made mirrors, but are comparable with the size of camera pixels.

4.4. *Muon rings*

We selected 'good muon' events to study the performance of our telescopes in the following criteria: 1) there is a cluster with enough number of adjacent hits in the image; 2) *arclength*, or ring radius times arc angle, is larger than 2 deg·rad; 3) the ring fit is fairly good (small χ^2). The curvature distribution for these muon events shows a clear peak around $1/1.2°$, which corresponds to local muons. We studied these muon events in detail and they are used to understand the performance of our telescopes [14].

4.5. *Global trigger system*

Since March 2004 we introduced a "global" trigger system which requires real-time coincidence between telescopes. For each telescope, when a local trigger is generated, data from the electronics are kept for 5 μs and at the same time the local trigger is sent to the stereo trigger system through an optical link. In the stereo trigger system, determination of stereo events is done by requiring any two or three local trigger signals coinciding with each other for at least 10 ns within a 650 ns time window considering the geometrical time delay which depends on the arrival direction of Cherenkov light. When a stereo event is detected, a global trigger signal is generated, and the event number and timing information are recorded. Then, the generated global trigger signal and event number are sent to each telescope. At each telescope, when the global trigger signal is received within 5 μs, ADC and TDC data are read out and recorded. If not, the data for that event are cleared, and the electronics and local trigger system are reset for a next event. Using this trigger system, local muon signal, which is evident in the length/size distribution taken in local triggers, disappear and we can reduce the local trigger threshold level in order to decrease the energy threshold.

5. Preliminary results from CANGAROO-III

5.1. *The Crab nebula*

From Woomera, the Crab nebula can be observed only at large zenith angles ($> 53°$). For stereo observations, The threshold energy of T1 is higher than other telescopes and thus we used the newer three telescopes for analysis. Because of the geometrical arrangement of the array, the effective baseline for large zenith angle observations becomes short which makes stereo reconstruction of images difficult.

To overcome the unfortunate situation described above, we developed new analysis methods. To avoid the increased uncertainty of the intersection points, we introduced a new parameter, "IP distance" ($D_{\rm IP}$), which is defined as the distance between the intersection point and centroid of images. Then we searched best intersection points which minimizes the image widths and the difference between distance and $D_{\rm IP}$. This results in better angular resolution as seen in the θ^2 distribution in Monte Carlo simulations, where θ is the space angle between the source direction and the reconstructed arrival direction.

We observed the Crab nebula for 18.5 hours in December 2003 in so-called wobble mode, changing the pointing directions $\pm 0.5°$ in declination apart from the target every 20 minutes.

In addition to the conventional square cuts method using image parameters to enhance gamma-ray fractions, we applied two different analyses: the Likelihood method [15,4] and the Fisher Discriminant method. In the latter method, effectiveness of the parameters for the gamma-ray-like event selection is evaluated using the simulation, and we can optimize the weights of the parameters in estimating the probability of gamma-ray-like events.

Finally we obtained the spectrum of the Crab nebula in the energy range from 2 to 20 TeV, which is consistent within the statistical and systematic errors with other measurements [16,14].

5.2. *Cen A*

Centaurus A is a well-known radio galaxy and the nearest ($z = 0.00183$) active galactic nucleus. Although its jet axis may be inclined from our line-of-sight, its proximity may overcome the misalignment and it could be detectable in the TeV region. Indeed, it was reported as a TeV gamma-ray source about thirty years ago when it was at a high state [17]. We observed Cen A for 1,047 min. and 882 min. for T2-T3 and T2-T4 pairs, respectively, in 2004 March. The stereo-mode analysis in each pair did not show a gamma-ray signal in the signal region ($\theta^2 < 0.047°$) and we obtained an upper limit of 3.2×10^{-12} cm^{-2}s^{-1} for $E_\gamma > 530$ GeV (2σ level) [18].

5.3. *The Galactic diffuse emission*

The Galactic plane is the strongest source of gamma-rays in the GeV region. The dominant mechanism of this emission is believed to be the nuclear gamma-rays via neutral pion decays produced in collision of cosmic-rays with interstellar medium. The EGRET results showed harder ($\propto E^{-2.4\sim2.5}$) spectrum expected from the cosmic-ray origin ($\propto E^{-2.7}$) [19], and if this spectrum extends to the TeV region, this Galactic diffuse emission could be detectable with Cherenkov telescopes. In 2004 we observed two local maximum regions of the diffuse Galactic emission model developed by the EGRET team [19]: $(\ell, b) = (-19.5°, 0.0°)$ and $(+13.0°, 0.0°)$. For the former point, $b = \pm3.0°$ were used as off-source regions, and the latter point was observed in long ON-OFF mode. ON-source observations of 10.6 hours ($\ell = -19.5°$, T2-T3), 6.3 hours ($\ell = -19.5°$, T2-T4), 8.0 hours ($\ell = +13°$, T2-T3), and 8.5 hours ($\ell = +13.5°$, T2-T4) were analyzed, but we could not find a gamma-ray signal associated with the Galactic disk in both regions. Assuming that the gamma-ray spectrum shows a single power-law between a few GeV and 600 GeV, conservative upper limits to the spectral index were obtained: -2.17 and -2.12 (preliminary) for each region, respectively [20].

5.4. *Supernova remnant RX J0852.0-4622*

We applied the Fisher discriminant method to the stereo data for RX J0852.0-4622 observed in January and February 2004 using T2 and T3 taken in the wobble mode for 2,197 minutes in total. We used the north-west rim as a target point in the wobble mode. After the coarse selections, 1,204 minutes data was available. For the Fisher discriminant, we used four image parameters, lengths and widths, determined with each telescope independently. Finally the gamma-ray events was extracted by comparing the Fisher discriminant values between the SNR region and the background region. The excess count map is shown in Figure 1. Te region inside the solid arcs shows the maximum acceptance region, which is an overlap of the two field-of-views in the wobble mode. Also the one-degree arc from the SNR center is indicated by the dotted line. The strong gamma-ray emission from the NW rim is obviously seen, which was first reported by CANGAROO-II. This maximum acceptance region covers about a half of the whole SNR, and the integrated flux above 0.81 TeV is about 60% of the H.E.S.S. result [10], which value is reasonable considering our coverage of the SNR [21,22].

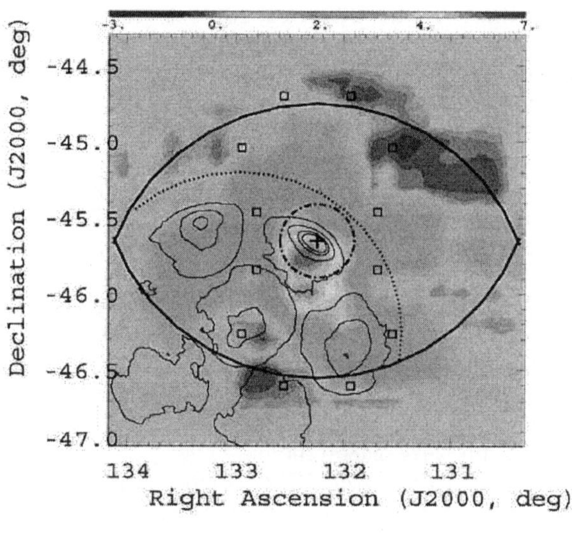

Figure 1. Excess event map around the SNR RX J0852.0-4622 obtained from the CANGAROO-III stereo observations (preliminary).

5.5. *Other sources*

We took fair amount of stereo data on the pulsar PSR 1706-44 and the supernova remnant SN1006 which we claimed to be TeV gamma-ray sources with CANGAROO-I. Preliminary analysis of both objects did not show any peaks in the θ^2 distribution [21]. (For SN1006 we take the northwest rim point where TeV gamma ray emission was reported by CANGAROO-I as the emission point.) Further analysis is underway so that we can check the possible extended emission.

6. Summary

We have been carrying out stereo observations of sub-TeV gamma-rays with four telescopes since March 2004. Preliminary results from stereo observations were presented here: further results are coming soon.

References

1. T. Tanimori et al., in *Proc. 26th ICRC (Utah)* (University of Uta, Utah), **5**, 203–206 (1999).
2. H. Kubo et al., *New Astronomy Reviews* **48**, 323–329 (2004).
3. A. Kawachi et al., *Astropart. Phys.* **14**, 261–269 (2001).
4. R. Enomoto et al., *Nature* **416**, 823–826 (2002).
5. F.A. Aharonian et al., *Nature* **432**, 75–77 (2004).
6. K. Tsuchiya et al., *Astrophys. J.* **606**, L115–L118 (2004).
7. K. Kosack et al., *Astrophys. J.* **608**, 97 (2004).
8. F. Aharonian et al., *Astron. Astrophys.* **425**, L13–L17 (2004).
9. H. Katagiri et al., *Astrophys. J.* **619**, L163–L165 (2005).
10. F. Aharonian et al., *Astron. Astrophys.* **437**, L7–L10 (2005).
11. R. Enomoto et al., in *Proc. 28th ICRC (Tsukuba)* (Universal Academy Press, Tokyo), pp. 2807–2810 (2003) and references therein.
12. K. Nishijima et al., in *Proc. 29th ICRC (Pune)*, in press (2005).
13. M. Ohishi et al., in *Proc. 28th ICRC (Tsukuba)* (Universal Academy Press, Tokyo), pp. 2855-2858 (2003).
14. R. Enomoto et al., submitted for publication (2005).
15. R. Enomoto et al., in *Proc. 27th ICRC (Hamburg)* (Copernicus Gesellshaft, Germany), Vol.5, pp. 2477–2480 (2001).
16. T. Nakamori et al., in *Proc. 29th ICRC (Pune)*, in press (2005).
17. J.E. Grindlay et al., *Astrophys. J.* **201**, 82–89 (1975).
18. S. Kabuki et al., in *Proc. 29th ICRC (Pune)*, in press (2005).
19. S.D. Hunter et al., *Astrophys. J.* **481**, 205–240 (1997).
20. M. Ohishi et al., in *Proc. 29th ICRC (Pune)*, in press (2005).
21. T. Tanimori et al., in *Proc. 29th ICRC (Pune)*, in press (2005).
22. R. Enomoto et al., to be submitted for publication (2005).

THE STATUS OF VERITAS

M. K. DANIEL ON BEHALF OF THE VERITAS COLLABORATION*

Department of Experimental Physics,
University College Dublin,
Belfield,
Dublin 4,
Ireland.
E-mail: michael.daniel@ucd.ie

The Very Energy Radiation Imaging Telescope Array System (VERITAS) in its first phase of operation will consist of an array of 4 Imaging Atmospheric Cherenkov Telescopes (IACTs) located at a high, dark site in Southern Arizona. The full array is expected to see first light in October 2006. In February of 2005 the first VERITAS telescope achieved first light at a temporary location near to the final site. This paper summarises the status of the VERITAS instruments as of May 2005.

1. Introduction

Phase-I of the VERITAS project will see an array of 4 IACTs sited at Horseshoe Canyon on Kitt Peak (111.6°W, 31.96°N, 1778 m a.s.l.) in a 'Mercedes' star configuration of 80 m spacing between telescopes. The first of the VERITAS telescopes have been operating at a temporary site located at the Smithsonian Institution's Administrative Complex at Mt. Hopkins (110.9°W, 31.69°N, $\sim$ 1300 m a.s.l.) in an engineering phase of operations. Even in this debugging period the telescope has achieved very reasonable scientific results, with Telescope-1 (T1) achieving first light in February of 2005. These results will be outlined in section 3; firstly, though, the technical aspects of the telescopes will be summarised.

*See `http://veritas.sao.arizona.edu/veritas_members.html` for a list of collaboration members

2. Hardware

2.1. *Mechanical/Optical Performance*

The VERITAS telescopes are of an f/1 Davies-Cotton design with a 12 m aperture. The mechanical structure consists of a commercial alt-alzimuth mount (model PG-4000) manufactured by RPM-PSI (Northridge, California) and a steel optical support structure (OSS) custom designed by M3 Engineering (Tucson, Arizona) and fabricated by Amber Steel (Chandler, Arizona). The telescope can be seen in the lefthand picture of figure 1. The tracking of the mount is measured to be accurate to $< 0.01°$ and it has a slewing speed of $\sim 0.3\,°/\mathrm{s}$. A slightly modified drive system has been tested with a slew speed of $1\,°/\mathrm{s}$; T1 will be upgraded to this new system and the remaining telescopes will have it installed as standard.

The reflector is segmented: consisting of 350 individual hexagonal mirror facets of $0.332\,\mathrm{m}^2$ each, providing a total mirror area of $\sim 100\,\mathrm{m}^2$. The facets are made of slumped and polished glass produced by D.O.T.I. (Austin, Texas), which are then coated with aluminium and anodized at a dedicated facility on-site. Reflectivity is $> 85\%$ from 280 to 450 nm. The mirror elements are well within specifications with a radius of curvature of $< R >= 23.92\,\mathrm{m} \pm 0.4\%$ and a blur circle diameter of $< D >= 4.9 \pm 1.9$ mm. The point spread function (PSF) of the telescope for the observations outlined in this paper was 0.09° at 80° elevation, which is well within the pixel size of the camera. Even so, the telescope alignment has subsequently been further refined and the current PSF is 0.06°.

Figure 1. Pictured on the left, the first VERITAS telescope located at its temporary site; on the right is a close up of the 499 pixel camera.

2.2. *Camera/Acquisition Electronics*

The camera consists of 499 Photonis XP2970/02 photomultiplier tubes (PMTs), with a 0.15° spacing giving a total field of view of 3.5°. A close up of the camera can be seen in the righthand picture of figure 1. The PMTs are run at a gain of 2×10^5. The signal from the PMTs is then further boosted through a high bandwidth pre-amplifier located at the base of the PMT before being sent through $\sim$ 50 m of RG59 stranded cable to the trigger and data acquisition electronics. Located in the camera is a current monitoring system for measuring the PMT anode currents. Currents for a bright field (such as the Crab Nebula) are typically 6 μA and for a dark field (such as Mrk 421) around 3 μA at the present gain level of the PMTs. These correspond to a night sky photo-electron background rate of 100-200 MHz per PMT.

The trigger is a 3 level system: the first level (L1) being a programmable constant fraction discriminator (CFD); the output of which is then passed to the second level (L2) pattern recognition system which is pre-programmed to recognise triggers resembling true Cherenkov light flashes; the final stage (L3) will be an array level trigger to discriminate coincident multiple telescope triggers. The plot on the right of figure 2 shows the trigger rate as a function of CFD threshold for two pattern trigger configurations for T1: a 3-fold adjacent pixel pattern and a 4-fold adjacent pixel pattern. The observations detailed here have been performed with a conservative CFD threshold of 6.7 photo-electrons and a 3-fold adjacent pixel pattern trigger, which corresponds to a telescope raw trigger rate of $\sim$ 150 Hz.

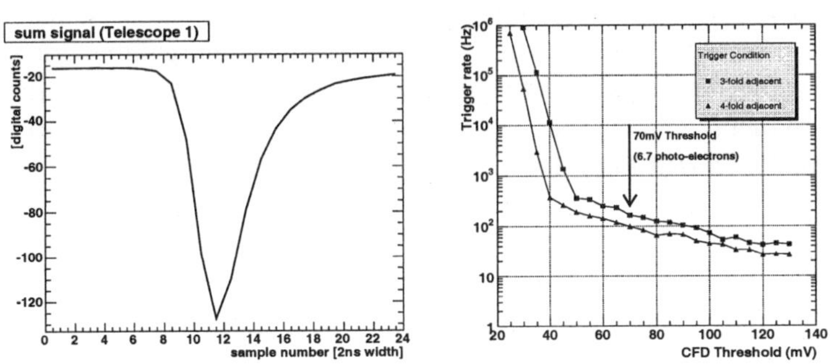

Figure 2. On the left: an FADC trace for a laser pulse, averaged over all pixels. On the right: the trigger rate as a function of CFD threshold for two different pattern trigger configurations.

The PMT signals for triggered events are digitised using custom designed 500 MHz Flash-ADC (FADC) boards. The default for the system follows a high gain path, but if the 8 bit dynamic range is exceeded an analogue switch moves to a delayed low gain channel instead - extending the dynamic range for a single 2 ns sample from 256 to 1500 digital counts. For the conservative trigger threshold presently used the system has a deadtime of $\sim 10\%$. When running as a full array the threshold will be reduced to allow better sampling of low energy gamma rays, meaning the individual trigger rate for an a telescope could be as high as ~ 1 kHz. In order that the system does not become deadtime dominated from reading out multiple channels of data at such high rates a zero suppression scheme will be implemented: channels that do not have a signal contained within them (to some pre-defined level) will not be read into the data-stream for that event. The electronics noise for the system is low – with a sample-to-sample variance of ~ 0.5 d.c. and an event-to-event variance of ~ 1.5 d.c. for a 10 sample integration window. Having an FADC system also allows us to minimise the charge integration gate and further promote the signal/noise ratio.

The telescope is equipped with a nitrogen dye laser, the light from which is transmitted via optical fibre and dispensed through an opal diffuser so that the camera can be illuminated by a uniform light source to make relative gain and timing corrections on the FADC traces. The lefthand plot of figure 2 shows the average trace recorded by one of these laser flashes. The laser is also equipped with a set of variable neutral density filters that allows the flashes to be attenuated over a large dynamic range to characterise the response of the system to large and small signals. Also located in the camera is a charge injection system, which allows us to inject a series of known artificial pulses, mimicking those of an actual Cherenkov pulse, into the electronics onwards from the pre-amplifier. The charge injection system will allow the testing and debugging of the system even during daylight, thereby reducing the engineering downtime for periods when observations can be made - a real boon for a technique with a duty cycle of $\sim 10\%$.

3. Observations and Data Analysis

As mentioned previously, this early phase of operation for T1 has largely been an engineering shakedown. Routine observations with T1 began in February 2005, in time to collect small datasets on the Crab Nebula (the standard candle for the field), Mrk 421 and Mrk 501 at high angles of elevation. Observing these known sources is good for verifying and quantifying the capability of this instrument and, consequently, the final array. Observations were taken in the standard ON-OFF pair mode and events were parameterised and selected according to the standard method used for the Whipple 10 m telescope [4]. The addition of FADCs into the data acquisition chain means we can also study the time profile of Cherenkov showers and use this information as an extra discriminant to distinguish between hadronic and electromagnetically induced air showers. Studies are ongoing to evaluate how powerful this supplementary information will be.

A full simulations chain has been developed [3] the results of which show good agreement with actual data: both in terms of image parameter distributions, see figure 3, and in terms of raw trigger rates. For a Crab like spectrum [2] this results in T1 having a threshold energy† of $\sim$ 150 GeV, giving a gamma-ray trigger rate of 22 γ/minute. Since the muon background [5] for a single telescope is very large some hard imaging cuts are made on single telescope data in order to maximise the significance of a signal, these in turn push up the threshold energy of T1 to $\sim$ 370 GeV.

3.1. *Crab Nebula*

Due to its steady signal the Crab nebula is the standard candle for TeV astronomy, therefore observations of this object are useful for gauging the performance of T1 and comparing its performance to current and previous instruments. The results shown in figure 4 are for a total of 3.9 hours on source observations taken at high elevation. This resulted in a 19.4 σ detection at a rate of $\sim 2.5\gamma$/min, which gives an idea of how harsh the cuts are that maximise significance. This shows that T1 has a sensitivity of 10 $\sigma/\sqrt{\text{hour}}$ for a Crab-like source, demonstrating the improvement in performance over its predecessor, the Whipple 10 m IACT [1].

3.2. *Mrk 421*

Mrk 421 is a Blazar at a redshift of $z \sim 0.031$. This is a highly variable source most easily seen when it is in a high flaring state, but close enough that it can still be observed in its quiescent phase. Figure 5 shows the resultant alpha plot and 2-d significance map for 13.1 hours on source observations, which are significant at the 12.5 σ level.

---

†defined as the energy that the peak trigger rate occurs for a Crab-like spectrum.

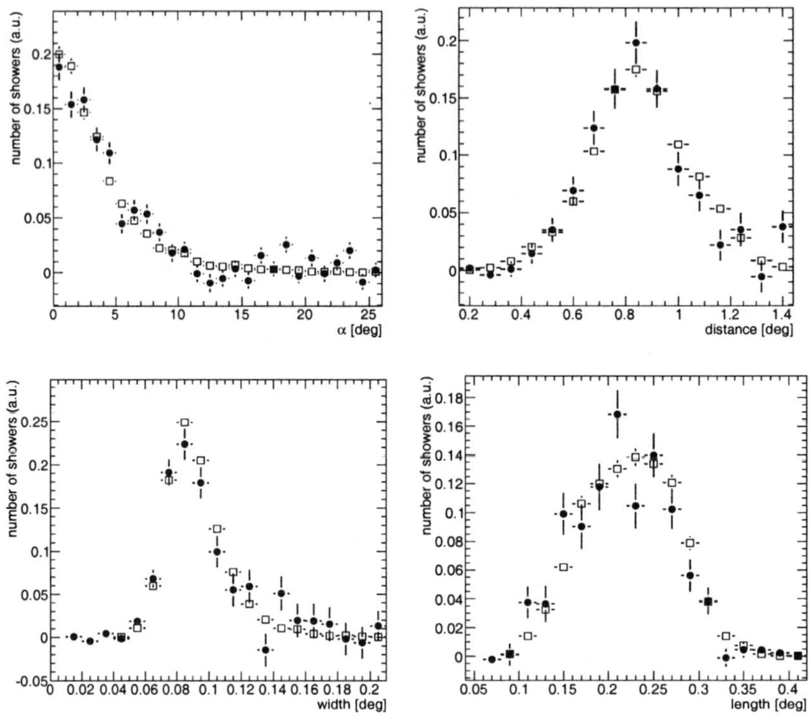

Figure 3. Image parameter distributions for observational data (filled circle) and Monte Carlo simulations (square symbol).

3.3. *Mrk 501*

Mrk 501 is another Blazar, at a similar distance to Mrk 421, with a redshift of $z \sim 0.034$. Again this a highly variable source, that was not in a high flux state at the time of observations, but still yielded a $7.1\,\sigma$ significance for 6.7 hours of on source observation time. Figure 6 shows the alpha plot and 2-d significance map for these observations.

4. Summary

The first of the VERITAS telescopes has been operating since February 2005 at a temporary site. Despite this site being at a lower elevation and having a brighter background light the telescope has met all technical specifications and is demonstrably an improvement over its predecessor. Even in this early, largely engineering, phase of operations T1 has shown good

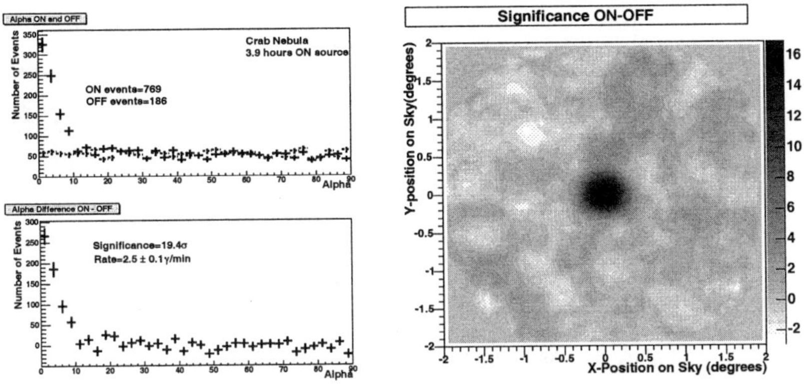

Figure 4. Results from T1 observations of the Crab nebula. On the left is the alpha plot, a gamma-ray source showing up as an excess of events at low alpha; on the right is a 2-d siginificance map of the source position in the field of view (Smoothing has been applied so adjacent bins are not independent).

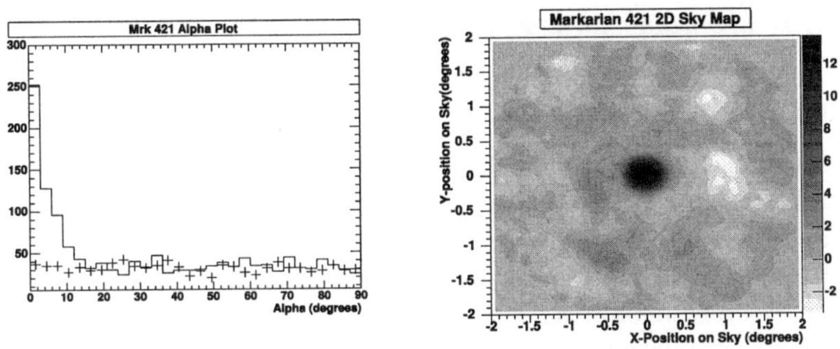

Figure 5. As figure 4, but for the Mrk 421 observations.

detections of a number of sources.

Due to a temporary delay in access to the final site at Kitt Peak the second of the VERITAS telescopes (T2) will also be temporarily installed at the Mt. Hopkins site, at a distance of 85 m from T1 in late 2005. Operating these two telescopes together will allow us to test the array trigger, which should eliminate the muon background that a single telescope suffers from and substantially increase the sensitivity of the system. Further improvements foreseen for the system will be the addition of light cones, which will decrease the amount of dead space between the pixels in the camera and increase the photon collection area. Two designs of light cones have been

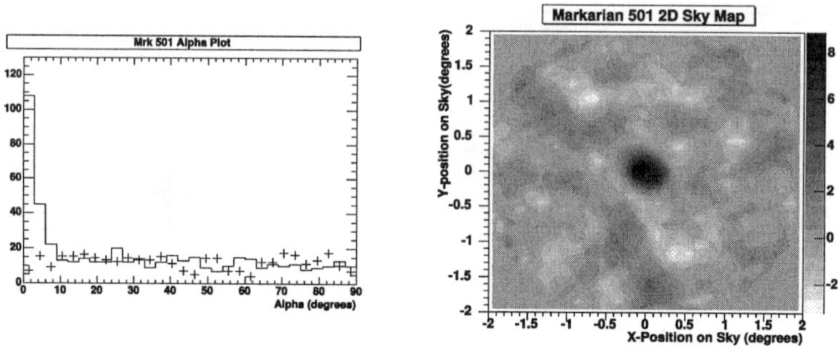

Figure 6. As figure 4, but for the Mrk 501 observations.

tested and a decision will be madc latcr in thc ycar as to which design shall be used in the final system.

The major mechanical components of all four telescopes have now been delivered and most of the infrastructure for the Kitt Peak site is in place. Once site access has been re-instated come January 2006 the full 4 telescope array is on schedule for first light in October 2006.

Acknowledgements

The technical assistance of E. Little and E. Roache are gratefully acknowledged in the running of the telescopes. VERITAS is funded in the United States by the *Department of Energy, National Science Foundation* and *Smithsonian Institution*; by *Science Foundation Ireland*; the *Natural Sciences and Engineering Research Council* of Canada; and the *Particle Physics and Astronomy Research Council* in the U.K.

References

1. Finley, J. P., et al., *Proceedings of the 27th ICRC (OG sessions)*, 2827 (2001).
2. Hillas, A. M., et al., *ApJ* 503, 744 (1998).
3. Maier, G., et al., *Proceedings of the 29th ICRC (OG sessions)*, (2005).
4. Reynolds, P. T., et al., *ApJ* 404, 206 (1993).
5. Vacanti. G., et al., *Astropart. Phys.* 2, 1 (1994).

GAMMA RAY BURSTS: RECENT RESULTS OBTAINED BY THE SWIFT MISSION

GUIDO CHINCARINI
ON BEHALF OF THE SWIFT TEAM
Università degli Studi di Milano Bicocca & Osservatorio Astronomico di Brera
Via E. Bianchi 46, Merate (LC), Italy

In the introduction we give the main characteristics of the Swift mission outlining how the design was driven by the science goals and by the heritage we got from the Italian – Dutch satellite Beppo SAX. We show some of the new characteristics of the X – ray light curves that became evident soon after we obtained the first set of data. In addition to the early phase steep slope afterglow discovered by Swift, we discuss the frequently observed GRB flares and the first localization of a short burst.

1. Introduction

With the launch of the Swift satellite, Gehrels et al. [1] a new era started not only for the study of the Gamma Ray Bursts but also for various fields of Astrophysics covering stellar evolution, relativistic stars and merging, relativistic jets and cosmology. Indeed the brightness of the GRB events supports the hope to detect these objects at very high redshifts and near the re-ionization epoch soon after the Universe gets out of the Dark Age. The design of the mission was carried out with some of these goals in mind and to some extent tailored about the discoveries made by Beppo- SAX, Costa et al. [2], Van Paradijs et al. [3], Frail et al. [4].

The main ingredients of the mission design were multi-wavelength coverage, state of the art instrumentation with good resolution and sensitivity, high speed re-pointing capability of the spacecraft and a fast and reliable communication system. These tasks were satisfied by a payload composed by three instruments: The Burst Alert Telescope (BAT), the X ray Telescope (XRT) and the Ultraviolet Optical Telescope (UVOT), by the Malind ASI ground station and TDRSS.

1.1. *Burst Alert Telescope (BAT)*

BAT, sensitive in the energy range 15 – 150 keV for imaging and with non-coded response reaching 350 keV, monitors the sky with its large field of view of about 1.4 steradians (100° x 60°). Within few tens of seconds of detecting a burst the on board computer will calculate the initial position with an accuracy of better than 4 minutes of arc and decide, after accounting also for all the pointing constraints, whether the burst merits a spacecraft slew. Assuming the burst is worthy, it also sends the position to the spacecraft for pointing the object and setting it on axis of XRT and UVOT. In various cases we were able to begin observations with the Narrow Field Instruments (NFI) within less than 60 seconds since the BAT alert.

1.2. *X ray Telescope (XRT)*

XRT is sensitive in the energy band 0.2 – 10 keV, has an effective area of 110 cm^2 and has the capability to estimate an accurate position, accuracy better than 5", in less than 100 seconds. Details of this instrument can be found in Burrows et al. [5]. An important characteristic of XRT is the capability to switch automatically between different modes of operations in order to have the capability to deliver accurate photometry of the source over a range of about seven order of magnitude (from 2 10^{-14} erg cm^{-2} s^{-1} to about 8 10^{-7} erg cm^{-2} s^{-1}). This is a requirement for observing bright and rapidly fading sources. Depending on the flux (count rate) of the source, therefore, the instrument switch from Image mode (IM) that is used at the very beginning (w/o spectral information) to determine the source position, to either Photo Diode mode (PD), used for very bright sources with a time resolution of 0.14 ms and no imaging information or Window Timing mode (WT) with a 1.8 ms resolution and space resolution in only one axis. For faint objects the instrument uses the Photon Counting mode (PC) with a time resolution of 2.5 seconds. This is the standard XRT mode. The quality of the X ray telescope is really excellent and on fields in which we did follow a burst for a long time we discovered we can reach a sensitivity limit of about 3 10^{-16} erg cm^{-2} s^{-1} (0.5 – 2 keV). These characteristics coupled to a good space resolution (18 arcsec HEW at 1.5 keV) over a 23 arcmin gave us the possibility of new discoveries in the field.

It is also known, however, that we lost the Thermo-Electric Cooling (TEC). Consequently the temperature of the CCD rather than being kept at -100 °C as planned, is oscillating between -85 and -45 °C since it uses only the passive cooling. Tests and calibrations that were carried out after that lost demonstrated we did not loose any sensitivity or resolution. On the other hand we must compensate with a more careful planning of the observations, to keep the CCD as cold as possible, and with more sophisticated calibrations and analysis.

1.3. *Ultraviolet Optical Telescope (UVOT)*

UVOT, Mason et al. [6], is a 30 cm aperture Ritchey – Chretien telescope with a 17x 17 arcmin FOV. It uses a CCD with a set of filters and grisms working in the wavelength range 1700 – 6500 Å with a space resolution of 0.5"x 0.5" on the sky. It can reach a position accuracy of 0.3 arcsec and a limiting magnitude (5 σ accuracy) B= 24.0 in 1000 s. It is essential in order to have a statistical significant sample not affected by the weather conditions and different size telescopes.

1.4. *The Transmission network*

Fast communication is essential for monitoring the observations of rapidly varying events and for alerting the ground based telescopes or other space facilities as soon as possible. To this end we use the TDRSS for immediate down and up loading of data and commands and the Italian ASI ground segment at Malindi for all ordinary operations.

2. Observations – First light

BAT performed quite well since the very beginning delivering excellent light curves in the 15 – 150 energy band and up to 350 keV. ùThe first image of XRT, Cas A, is reproduced in Figure 1 next to an image obtained with the XMM telescope for comparison. The space resolution is excellent.

3. Early science results

This Cividale meeting came at a very interesting time because we had the possibility to show the discovery of the main characteristics of the light curves observed by XRT and the accurate localization of the first short burst, GRB050509. These results, at the time of writing, have been confirmed with the detailed analysis of many XRT light curve and with the discovery of the short burst GRB050709 (HETE) and GRB050724. In addition Swift discovered the high z GRB050904 (z = 6.29) long after the meeting and this we will not discuss in this proceedings.

3.1. *XRT light curves*

The analysis of the following light curves, GRB011121, GRB050126, GRB050219A, GRB050315 and GRB050319, showed since the very beginning the main characteristics that have been described in a more organic way by Chincarini et al. [6] and by Nousek et al. [7].

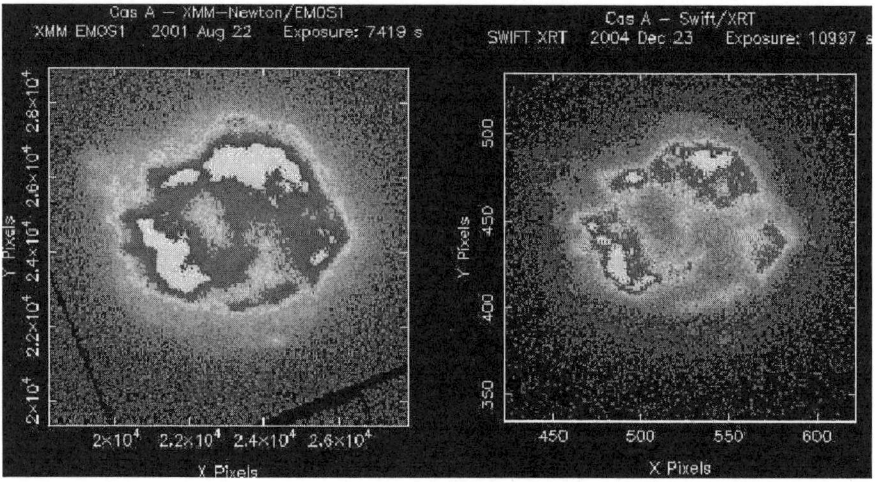

Figure 1. Left: Cas A obtained by XMM on August 2001 with an integration of 7419 seconds. Right: Cas A observed by XRT on December 2004, during the activation phase, with an exposure of 10997 seconds.

It is clear from Figure 2 that the basic morphology is given, soon after the prompt emission, by a fast decay of the afterglow, see also Tagliaferri et al. [8], followed by a milder decay steeping down after a while. This will define the type 1 light curve. In one case, GRB050319 in Figure 2, the initial slope is much higher (about -6) but that is almost certainly due to an erroneous BAT trigger. Counting the time (t_0) from the beginning of the second BAT burst (in this GRB the prompt emission is composed by two bursts separated by about 137 seconds) the light curve of GRB050319 is exactly superimposed to the light curve of GRB050315. A type 2 light curve seems also to occur. This starts off with a mild slope soon after the prompt emission to steep down later on, see also Figure 3 in Chincarini et al.. The type 1 light curve is shown in Figure 2 by the dashed dotted line that represents the mean of the various type 1 light curves seen in that plot. GRB 050401 is on the contrary the prototype of a type 2 light curve.

The X-ray spectrum remains unchanged when measured before and after the break, GRB050319 being an exception, and it is very similar for the various burst with an energy index $\beta \sim 1.1$. The energy measured in the afterglow changes considerably for the various bursts and is in between 1.6% to 40% of the energy emitted by the prompt emission.

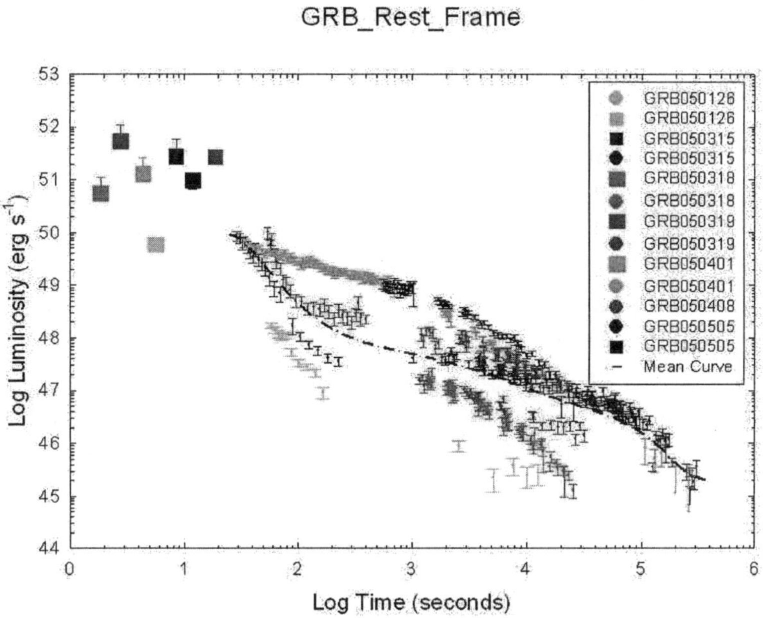

Figure 2. The first set of XRT light curves for which the redshift has been measured. There is a clear indication of continuity between the BAT and the XRT emission, for details see Chincarini et al.

The observations of GRB011121, GRB050219A, 050406 and 050502B showed very clearly that superimposed to these basic features we may observe flares, some of which are very energetic. A beautiful example of type 1 with flares is shown in Figure 3 (note however that this was obviously not shown at the time of the meeting) in which we wrote for convenience also the measured fluence.

Light curves with large flares have afterglow fluence (XRT) comparable to that of the prompt emission (BAT).

While the change of slope at the first break may need further modeling to be fully understood, the second break of the type 1 light curve (or the first of the type 2 light curve) occurs when the relativistic beaming angle

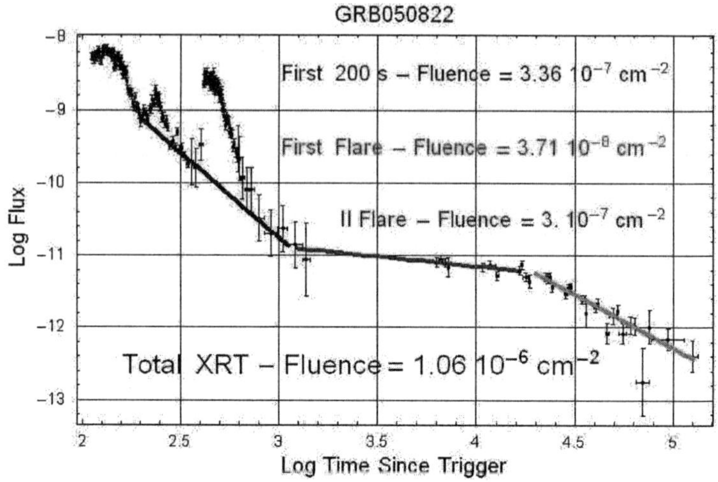

Figure 3. The XRT light curve illustrate properly all the characteristics described in the text. The fluence measured by BAT in the band 15 – 350 keV is 3.1 10^{-6} cm^{-2}.

reaches, because of the decreasing of Lorentz factor of the relativistic shells, the value of the jet angle, Figure 4.

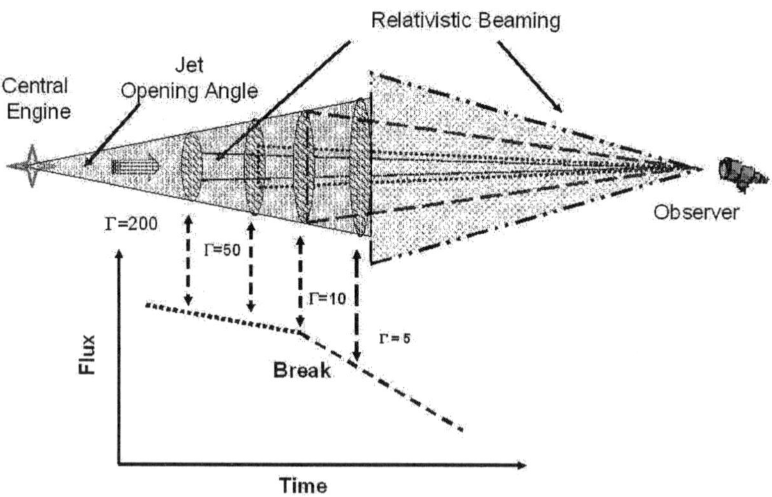

Figure 4. In this sketch the shell slow down from $\Gamma = 200$ to $\Gamma = 8$ and equals the Jet opening angle when $\Gamma = 10$ ($\theta = \Gamma^{-1}$). At that point the light curve shows a break and is steeping (the value of the Lorentz factors are only for illustration and are not real).

4. The localization of short gamma-ray bursts

The duration of gamma ray bursts is known to have a bimodal [9], short and long, distribution and BATSE showed that short bursts were harder [10]. Short bursts, duration less than 3 s, have been elusive for decades because of the difficulty in getting an accurate location soon after the alert. On May 5th BAT detected a burst, Gehrels et al. [11] (see error circle top left of Figure 5), with a fluence of 9.5 $\pm$ 2.5 10^{-9} erg cm^{-2} in the band 15 to 150 keV, right panel of Figure 5. Automatically the spacecraft re-pointed on the BAT coordinates and XRT was collecting data 62 s after the burst trigger.

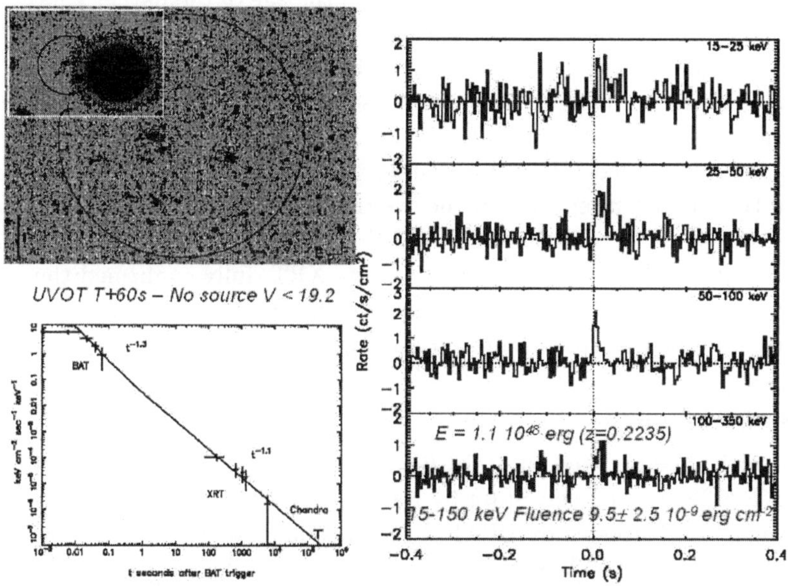

Figure 5. The first localization of a short gamma-ray burst. Top left: The field of the sky with the error circle of BAT and in the inset the E1 galaxy with the error circle of XRT. Left bottom: the XRT light curve including a late observation by Chandra. Right: the BAT light curve at different energies.

XRT collected 11 photons and gave the position with an error circle of only 9.8" (inset top left frame of Figure 5). The X-ray light curve (0.3 – 10 keV) (we can now show the complete light curve, bottom left panel of Figure 5, including the Chandra data) shows a classical decay power law with α = -1.1. We, and other groups, immediately imaged the field with medium large telescope (TNG,VLT, SUBARU, Keck etc.) and we noticed that the XRT error circle included the second brightest galaxy (E1) of the cluster of galaxies NSC J123610+295901. The cluster has a redshift z = 0.225. Within the XRT error circle we observed also various faint galaxies, likely at high redshift, and a somewhat brighter spiral.

We opted for the bright E1 elliptical as the host galaxy because of the extremely low probability that a random burst would have to be located next to a bright nearby galaxy. We were also guided in a small part by some of the possible theoretical predictions [12,13,14] (without having a prejudice however). Indeed we discussed merging of relativistic stars (NS-NS, BH-NS), SGR, peculiar collapsars etc. But the idea of NS - NS or NS - BH was very appealing and in agreement with the age of the parent galaxy and the outskirt location due to the rather large peculiar velocity gained by the binary system as a consequence of the kick off received after the supernova event. This interpretation agrees well also with the estimated energy release assuming the burst is at the distance of the galaxy, E_{iso}=1.1 10^{48} erg, and with the age needed for such a system to form and merge. Following this event that has been reported at the meeting in a preliminary form, the burst detected by HETE, GRB050709, and that detected by Swift, GRB050724, whose detailed light curve was followed by XRT, fully confirmed the above interpretation and we have now, and possibly more in the coming months, the data needed to constraint and understand the progenitor.

Acknowledgments

The work is supported in Italy by funding from ASI on contract number I/R/039/04, at Penn State by NASA contract NAS5-00136 and at the University of Leicester by PPARC contract number PPA/G/S/00524 and PPA/Z/S/2003/00507. We acknowledge in particular all those member of the Swift Team at large who made this mission possible. This goes from the building of the hardware, the writing of the software, the operation at the Mission Operation Centre and the performance of the ASI ground segment at Malindi, Kenya.

References

1. Gehrels, N., Chincarini, G., Giommi, P., et al. ApJ, 611, 1005
2. Costa, E., Frontera, F., Heise, J., et al. 1997, Nature, 387, 783
3. Van Paradijs, J., Groot, P.J., Galama, T. et al. 1997, Nature, 386, 686
4. Frail, D.A., Kulkarni, S.R., Nicastro, S.R. et al. 1997, Nature, 389, 261
5. Burrows, D. N., Hill, J. E., Nousek, J. A. et al. 2004 SPIE, 5165, 201

6. Chincarini, G., Moretti, A., Romano, P. et al. 2005, Astro – ph 0506453
7. Nousek, J.A., Kouveliotou, C., Grupe, T. et al. 2005, Astro – ph 0508332
8. Tagliaferri, G., Goad, M.R., Chincarini, G. et al. 2005, Nature, 436, 985
9. Cline, T.L. & Desai, U.D. 1974, Proceedings 9th ESALAB Symp. Noordwijk.
10. Kouveliotou, C., Meagan, C.A., Fishman, G.J. et al. 1993, ApJ 413, L101
11. Gehrels, N., Sarazin, C.L., O'Brien, P. et al. 2005, Nature, 437, 851
12. Paczynski, B., Acta Astronomica, 1991, 41, 257
13. Mochkovitch, R., Hernanz, M., Iserb, J. & Martin, Nature, 1993, 361, 236
14. Katz, J.I. & Cannel, L.M., 1996, Ap.J. 471, 915

FUNCTIONAL TESTS AND PERFORMANCE CHARACTERIZATION DURING THE ASSEMBLY PHASE OF THE MODULES OF THE AGILE SILICON TRACKER

M. BASSET *, E. VALLAZZA
INFN, sezione di Trieste

C. PONTONI
CIFS, Consorzio Interuniversitario per la Fisica Spaziale, Torino

L. FOGGETTA, M. PREST
INFN, sezione di Milano

G. BARBIELLINI, F. LIELLO, F. LONGO
Università degli Studi di Trieste and INFN, sezione di Trieste

AGILE (Extremely Light Imager for Gamma ray Astrophysics) is a small scientific satellite for the detection of γ-ray cosmic sources in the energy range 50 MeV-30 GeV with a large field of view (1/5 of the sky). The heart of the AGILE scientific instrument is a Silicon-Tungsten Tracker which consists of 12 planes with two layers of 16 single-sided, AC coupled, 410 μm thick, 9.5×9.5 cm^2 silicon detectors each. The detector readout pitch is 242 μm with a floating strip configuration. The readout chip is the TAA1, an analog-digital, low noise, self-triggering ASIC with analog readout. The total number of readout channels is about 37000.
After the prototype performance studies, the final flight version of the Tracker has been assembled in one year. A large number of tests (e.g. mechanical measurements, thermo-vacuum cycling tests, electrical and functional tests) have been performed on every Tracker element at every assembly step. We present an overview of the test procedures and of the performances of the flight model.

1. Introduction

AGILE is an ASI (Italian Space Agency) Small Scientific Mission, dedicated to high energy astrophysics. The AGILE scientific instrument is made of

*Corresponding author, address: INFN - Sezione di Trieste, Area di Ricerca, Padriciano 99, I-34012 Trieste, Italy; e-mail: marco.basset@ts.infn.it

three integrated detectors (fig. 1): the Silicon-Tungsten Tracker (ST), based on alternate layers of tungsten sheets (to convert the photons) and silicon microstrip detectors (to detect the electron positron pairs), which is able to detect and image photons in the energy range 50 MeV-30 GeV with excellent spatial and angular resolution, good timing capability and a very large field of view (about 1/5 of the sky); the SuperAgile detector (SA), based on a coded mask and silicon microstrip detectors, which is able to detect and image photons in the energy range 15-45 keV; the CsI MiniCalorimeter (MCal) which is able to detect and measure the energy of photons in the range 300 keV-200 MeV. Moreover the AntiCoincidence system (AC) made of plastic scintillators is placed around the instrument for background rejection.

A detailed description of the Silicon Tracker is given in [[1,2]]. The assembly and test of the flight model took about one year. In the following, after describing briefly the detector (Section 2), the assembly (Section 3), the test procedures and the obtained results (Section 4) are presented.

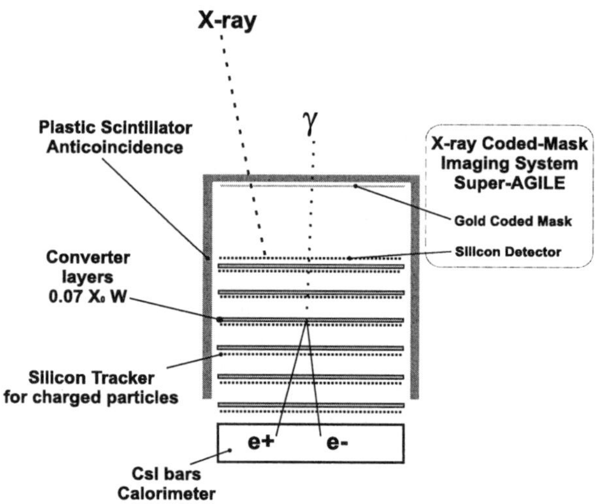

Figure 1. A sketch of the AGILE scientific instrument and its working principle.

2. The AGILE Silicon Tracker

The AGILE Tracker is made of 12 detection planes, each one consisting of two views of 16 silicon microstrip tiles (Hamamatsu), organized in 4 mod-

ules ("ladders") of 4 tiles. The two views are positioned orthogonally one with respect to the other in order to obtain a x-y imaging system and the planes are organized in 13 mechanical trays. The first 10 planes have a 245 μm (corresponding to about 0.07 X_0) thick tungsten layer for the photon conversion.
Each tile consists of a 9.5×9.5 cm^2 AC-coupled 410 μm thick 384 silicon strip detector. The 4 ladder tiles are connected in series strip by strip through 25 μm thick aluminium ultrasonic wire bondings. The strip physical pitch is 121 μm, while the readout one is 242 μm with one floating strip in order to reduce the number of electronics channels and thus the power consumption while maintaining a good spatial resolution.

The readout ASIC is the TAA1 (fig. 2b), a 128 channel, low noise, low

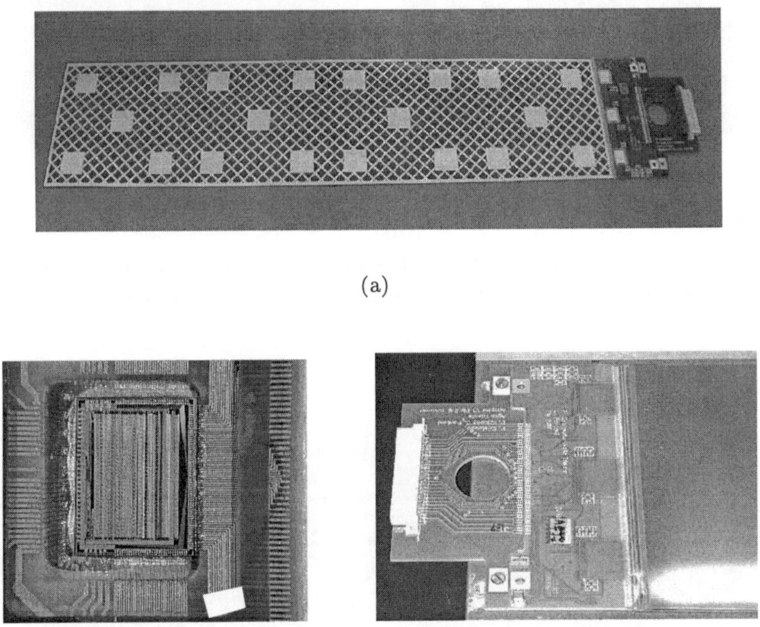

(a)

(b)

(c)

Figure 2. (a) Picture of the HDI PCB. The 38 cm long kapton cable for the bias is clearly visible; the HDI output is connected to the temporary adapter for the tests. (b) Picture of the TAA1 ASIC. (c) Detail of the HDI-ladder interface: the FR4 covers of the three ASICs are visible; the bondings between the HDI and the first silicon tile are covered by silicone glue which is contained by an epoxy dam.

power, self triggering ASIC designed by Ideas (Norway) and produced by AMS (Austria) with a 0.8 μm N-well BiCMOS, double poly, double metal on epitaxial layer technology. To limit the power consumption of the Tracker, the ASIC is operated in a very low power configuration (< 400 μW per channel). The die is 5.174×6.919 mm^2 and ~ 600 μm thick. The 128 input pads have a 100 μm pitch and have been bonded to the PCB lines with a 17.5 μm thick aluminium wire, while a 25 μm one has been used to bond the 41 output, control and power 200 μm pitch pads.

The HDI (High Density Interconnection) is the circuit that interconnects

(a) (b)

Figure 3. (a) A ladder on the CMM table during the mechanical measurements. (b) Glue dispenser during the preparation of the kapton layer for the silicon ladder.

(a) (b)

Figure 4. (a) A tray with the four ladders glued on one of the views. (b) A picture of the Tracker.

the ladder ASICs and the detectors to the readout electronics. It consists of a ten layer FR4 PCB (Printed Circuit Board) designed by INFN Trieste (Italy) and produced by ILFA GmbH (Hannover, Germany). One of the layers is a 38 cm long kapton foil extending outside the PCB (fig. 2a) which is used both as a mechanical support for the silicon detectors and to provide the bias to the detectors themselves.
Each tray (Oerlikon-Contraves, fig. 4a) is made of a 15 mm thick aluminium honeycomb core with a 0.5 mm carbon fiber layer per side. The tungsten foil is glued on to the bottom part of the first 10 trays.

3. The assembly steps

The prototype and flight model assemblies have been performed by Mipot S.p.A. (Cormons, Italy).
The assembly steps can be divided into four groups:

- Assembly of the passive components and of the ASICs on the HDIs (fig. 2). The following operations are performed on the PCBs, once they have passed the incoming inspection: tin soldering of the passive components (MIL-883 resistors and capacitors); gluing and bonding of a temporary adapter for the connection to the test equipment; gluing (Epotek H20E), bonding and conformal coating (RTV615) of the ASICs; gluing of the FR4 covers; final 72 hour test in a thermo-vacuum chamber with temperatures ranging from -30°C to +50°C; each cycle lasts 2 hours.
- Assembly of the silicon ladder. Four tiles are glued head-on (AV138) and the so obtained silicon ladder is glued on the HDI kapton (Epotek H20E on the gold dots for the elctrical connection to the bias voltage and DC3145 on the rest) (fig. 2c, fig. 3). The head-on gluing between the first silicon tile and the HDI is made with three segments of AV138 separated by RTV615, to avoid breaking caused by eventual thermal gradients. The strips are then bonded together between the four silicon tiles and between the HDI and the first silicon tile. The bondings are protected with an epoxy dam (2216) filled with RTV615 (fig. 2c).
- Tray assembly. Four ladders are glued first on the bottom side of a tray and then on the top view (fig. 4a). The adapters are taken away and the cable to connect the frontend to the readout electronics (MLC, MultiLayer Connection) is fixed with screws and glued. The HDI and the MLC are bonded with 25 μm Al wire and

the bondings are protected as in the silicon ladder case.

- The 13 trays are assembled together with the mechanical supports, obtaining the Silicon Tracker (fig. 4b).

As far as each step duration is concerned, the glue hardening requires one day and 8 hours at 50°C in case of epoxy glues.

4. The tests

Each assembly step required a set of tests which can be divided into:
Mechanical measurements. They have been performed with a CMM (Coordinate Measuring Machine) by Mitutoyo featuring a precision of the order of a few microns and two operation modes, a mechanical one with a "finger" touching the surface and an optical one with a video camera focusing on the point chosen by the operator. This second mode is fundamental when measuring the position of the silicon detectors on the tray. Each tray has been mapped before assembly to check the mechanical dimensions and the planarity. A dedicated procedure has been developed to verify the alignment of the single tiles in the ladder, the position of the ladder with respect to the HDI and its position once glued on the tray.

Electrical tests. Each HDI has been tested once the passive components

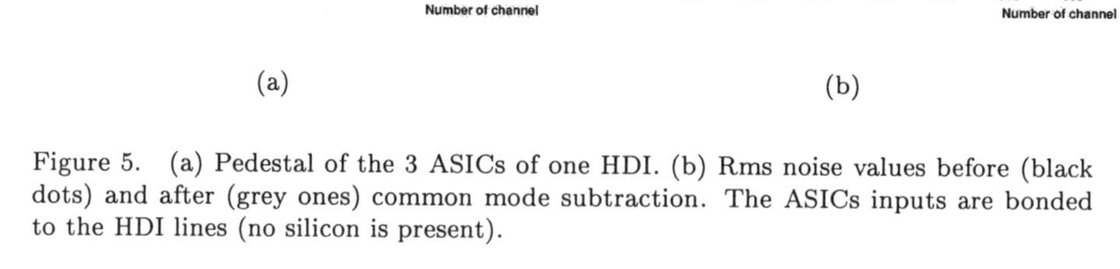

Figure 5. (a) Pedestal of the 3 ASICs of one HDI. (b) Rms noise values before (black dots) and after (grey ones) common mode subtraction. The ASICs inputs are bonded to the HDI lines (no silicon is present).

were assembled on the PCB to verify open and short circuits.
Current consumption tests on the ASICs. The HDI current absorption has been verified after every assembly step. ASIC breakdown due to latch-up phenomena occurs often at the beginning of the ASIC operation.

Pedestal and noise tests. This test has been performed at every assembly step. Pedestal and noise levels have been checked to find eventual problems in the PCB (open circuits due to the assembly procedure), in the bonding wires (disconnected wires or shorts) and in the ASICs. Long term pedestal acquisitions of more than 8 hours have been performed for each HDI/ladder to test its stability. Fig. 5a shows the pedestal of one of the HDIs while fig. 5b the rms noise before and after common mode subtraction. The ASICs inputs are bonded to the HDI pitch adapter lines (no silicon is present) which have different lengths resulting in the noise structure. The high noise values indicate the presence of a short.

Calibration tests. The TAA1 is provided with a calibration input line

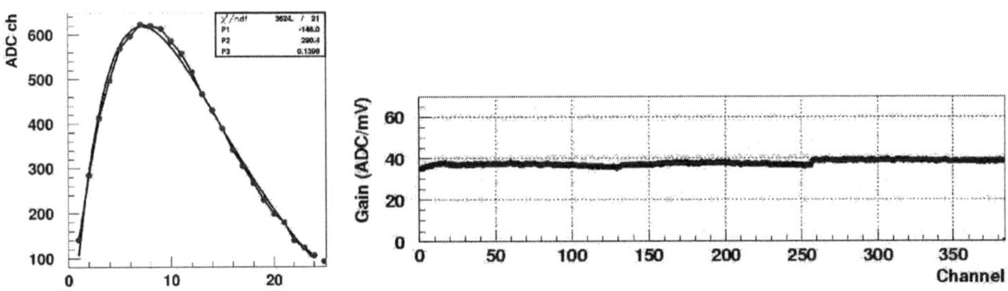

(a) (b)

Figure 6. (a) Analog shape of one TAA1 channel. Both the axis are in arbitrary units. (b) Gain in ADC/mV of the three ASICs of one HDI.

in order to send a given signal to one channel at a time. Using the hold signal, the output amplitude can be sampled at different times from the signal generation thus obatining the shape of the analog signal (fig. 6a) of each channel. This procedure has been applied to all channels with 3 different input signal in order to calibrate the ASICs response. A shorter procedure has been performed on each ASIC after each following assembly step. Fig. 6b shows the gain of three ASICs of one HDI as computed using the calibration procedure.

Tests on the digital part of the ASIC. The TAA1 is a self-triggering ASIC with a discriminator per channel. The possibility to disable/enable each channel has been tested at each assembly step. A dedicated procedure has been implemented to perform a threshold scan for each channel

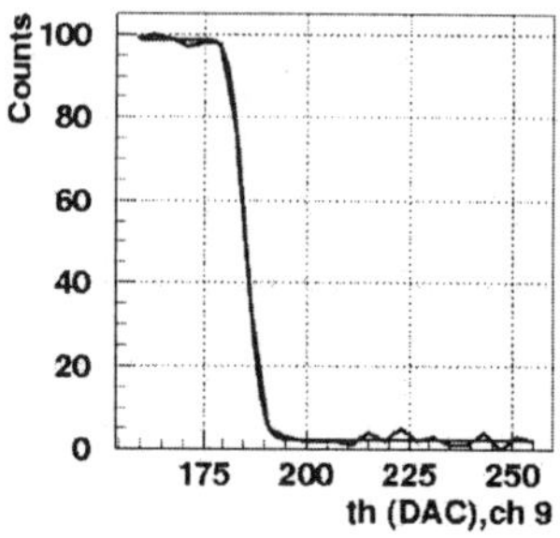

(a)

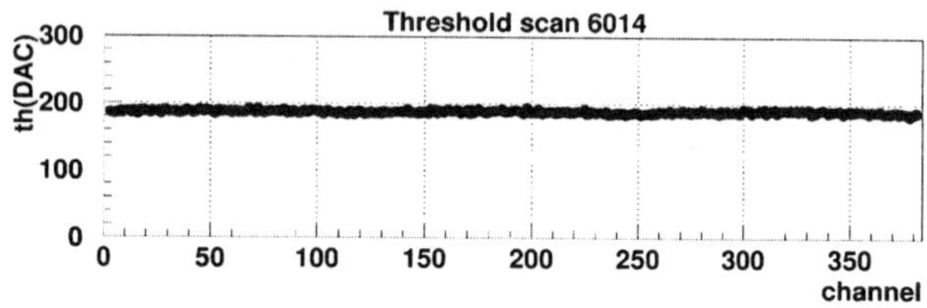

(b)

Figure 7. (a) Example of a threshold scan on a channel; the horizontal axis is in arbitrary units. (b) Calibration signal amplitude computed from the threshold scans of each channel.

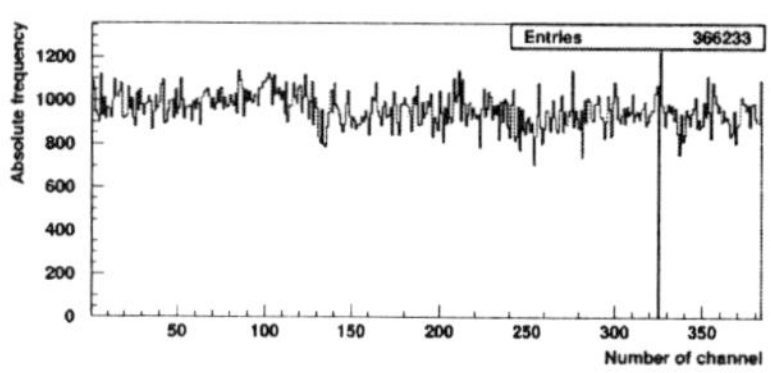

(a)

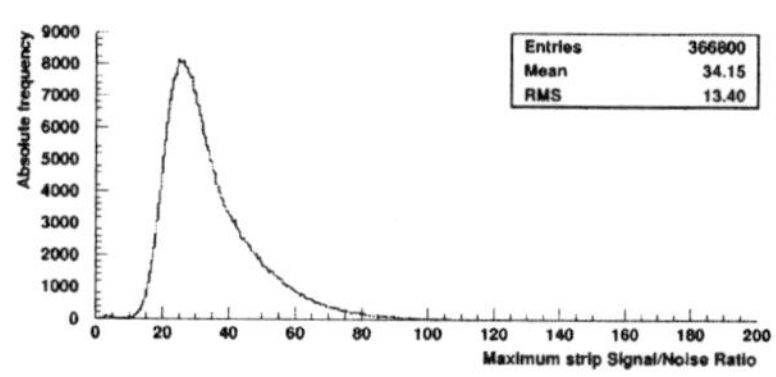

(b)

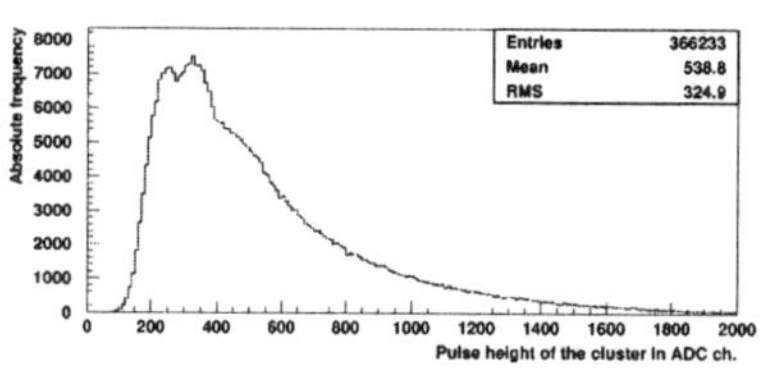

(c)

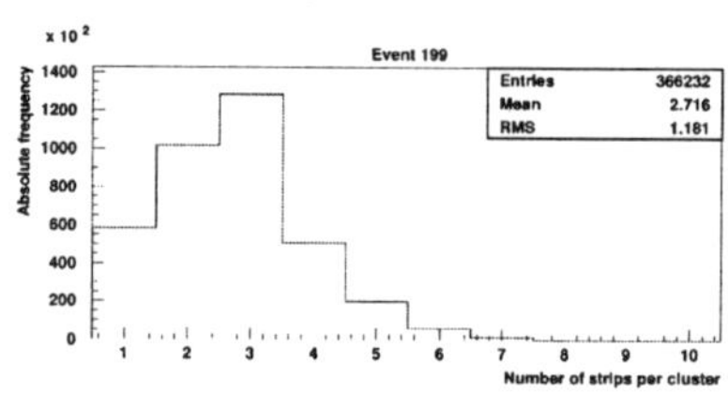

(d)

Figure 8. Typical cosmic ray results: (a) beam profile; (b) signal to noise ratio for the strip with the maximum signal; (c) cluster pulse height; (d) number of strips per cluster.

injecting a known signal in the calibration input. Fig. 7a shows a typical threshold scan while fig. 7b presents the response of the three HDI ASICs to a given calibration signal as computed from the threshold scans.

Functionality tests with cosmic rays. The 96 modules of the flight model have undergone a long term (a few days) cosmic ray test before and after the integration on the trays. Fig. 8 shows a few summary plots: the beam profile (a) where a dead channel is visible; the signal to noise ratio of the strip with the maximum signal in the event (b); the cluster pulse height in ADC units (c); the number of strips per cluster (d). A cluster is defined as the group of strips that have detected the particle crossing. The particle has been detected if at least one strip has a signal greater than 10 times the rms noise. The nearby strips are considered part of the cluster if their signal to noise ratio is greater than 5.

Shorter cosmic ray runs have been performed after every assembly test. In table 1 the number and type of defects of the flight model modules in terms of noisy/dead strips are listed.

Table 1. Defective channels (the total number is 36864). The second row presents the ranges of the ratio between the channel noise level and the mean noise level. The 5th column indicates the group of noisy channels due to the presence of defects in the silicon tiles, while the last column shows the number of channels that have to be disabled at the ASIC level to avoid non physical triggers.

	Dead	Low	Noisy	Very noisy	Si defects	Disab. channels
Ranges	$0.0 - 0.4$	$0.4 - 0.8$	$1.2 - 3.0$	> 3.0		
N. of channels	108	41	205	36	<15	~ 60

5. Conclusions

The flight model of the AGILE Silicon Tracker has been delivered for the AIV (Assembly, Integration and Verification) to Laben S.p.A. on the 1st of July 2005. The construction of the Tracker has required the equivalent of more than 10 man years of tests. The AGILE satellite will be launched during 2006 by the Indian launcher PSLV on an equatorial orbit at 550 km.

References

1. G. Barbiellini *et al.*, Nucl. Instr. and Meth. in Phys. Res. A 490, 146-158, 2002.
2. G. Barbiellini *et al.*, *The next generation of high-energy gamma-ray detectors for satellites: the AGILE Silicon Tracker*, AIP Conference Proceedings: 587, 754-758, 2001.

STATUS OF GLAST, THE GAMMA-RAY LARGE-AREA SPACE TELESCOPE

LEON ROCHESTER
ON BEHALF OF THE GLAST TEAM
Stanford Linear Accelerator Center, Stanford University, PO Box 20750, Stanford, California 94309, USA

GLAST is a satellite-based observatory consisting of the Large-Area Telescope (LAT), a modular 4x4-tower pair-conversion telescope with a field-of-view greater than 2 steradians, capable of measuring gamma-ray energies in the range 20 MeV to 300 GeV, and the GLAST Burst Monitor (GBM), a set of NaI and BGO detectors covering 8 steradians and sensitive to photons with energies between 10 keV and 25 MeV, allowing for correlative observations of transient events. The observatory is currently being constructed and is scheduled to be launched in August 2007.

1. Introduction

GLAST [1] is a satellite-based observatory consisting of two instruments: the Large-Area Telescope (LAT) and the GLAST Burst Monitor (GBM). Together they will provide for correlative measurements of gamma-ray phenomena in the energy range between 10 keV and 300 GeV. The observatory is scheduled to be launched in August 2007 for an expected five- to ten-year mission.

2. The GBM

The GLAST Burst Monitor [2] is similar in design to the highly successful BATSE instrument onboard the CGRO mission and will provide similar capabilities. It is a set of sodium iodide (NaI) and bismuth germanate (BGO) detectors mounted around the outside of the satellite. It has a sky coverage of 8 steradians, which enables prompt alerts to the LAT detector and to the ground of a region of interest, as well as autonomous repointing requests for events outside the current field-of-view of the LAT. The localization precision is expected to be better than 15° in the first several seconds, with an ultimate resolution of better than 1.5°. Its spectral response just overlaps that of the LAT, which allows for simultaneous spectral fitting. The assembly of the GBM is on schedule for delivery for spacecraft integration in January 2006.

3. The LAT

The Large-Area Telescope [3] is a pair-conversion telescope, which takes much of its basic design concept from its predecessor EGRET [4,5]. It is built of 16 identical modules, called towers, enclosed in an anti-coincidence detector (ACD) made of overlapping scintillator tiles, which are read out by optical fibers. Each tower consists of a tracker and a calorimeter. The tracker (TKR) is based on silicon-strip technology, and features 18 pairs of planes, measuring orthogonal coordinates, and separated by thin tungsten-alloy conversion foils. A calorimeter (CAL) made of a hodoscopic stack of thallium-doped cesium iodide [CsI(Tl)] crystals is situated below the tracker in each tower. Figure 1 shows a schematic representation of the LAT.

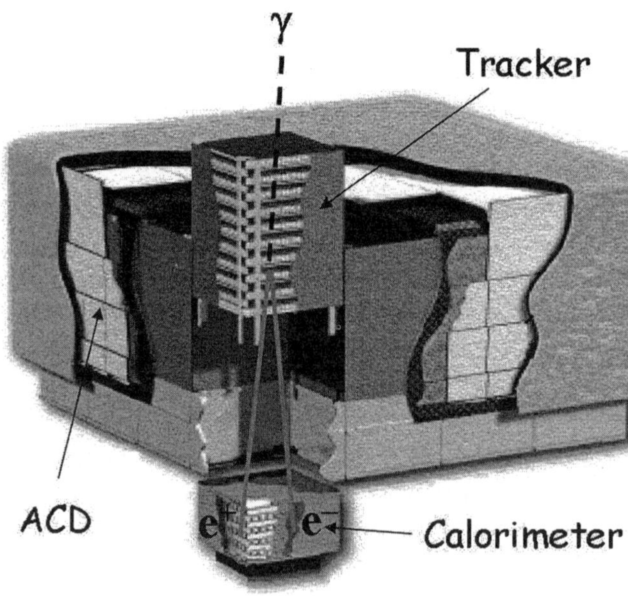

Figure 1. The Large Area Telescope is composed of sixteen modules, each containing a tracker and a calorimeter, all surrounded by a scintillator anticoincidence shield.

When a photon traverses the detector, there is some probability that it will convert or materialize into an electron-positron pair in the tracker, where the paths of the two daughter particles can be measured in the silicon-strip detectors. The electron and positron typically exit through the bottom of the trackers and produce an electromagnetic shower in the calorimeter modules.

For every high-energy photon impinging on the detector, there are several thousand cosmic rays. The role of the ACD is to aid in distinguishing the cosmic rays from the photons; photons will in general pass through the ACD undetected, whereas charged particles will produce scintillation light.

4. Comparison with EGRET

GLAST is a follow-on to the very successful EGRET detector which flew on the Compton Gamma-Ray Observatory [6,7] and took data between 1991 and 2000. The LAT was designed to improve on the performance of EGRET, in terms of field-of-view and resolution (and hence sensitivity), dead-time, background rejection, and reliability. Since the detectors are entirely solid-state, there are no consumables. The LAT silicon-strip detectors provide better measurement accuracy than the EGRET spark chambers, and significantly lower dead-time. EGRET was taller than it was wide, whereas the LAT is much more squat, which translates into a considerably larger field of view. Table 1 compares the performance of EGRET with that expected of the LAT.

Table 1. A comparison of the performance of EGRET and the LAT

	EGRET	LAT	Factor
Energy	20 MeV – 30 GeV	20 MeV – 300 GeV	10
Energy Resolution	~10%	~10%	1
Peak Effective Area	1500 cm^2	10,000 cm^2	6
Field of View	0.5 sr	2.4 sr	4
Sensitivity (1 year)	$\sim 10^{-7}\gamma cm^{-2}s^{-1}$	$3\ 10^{-9}\gamma cm^{-2}s^{-1}$	30
Source Localization	15'	< 0.5'	30
Dead-time	100 ms	< 50 μs	2000

5. Gamma-Ray Science

No instrument currently deployed explores the energy range accessible by the LAT. Also, the LAT is a survey instrument, sampling the entire sky over the course of a single day. These features position it to explore a wide range of phenomena.

5.1. *Galactic Diffuse Emission*

The LAT sensitivity and resolution will help us to resolve many more point sources than were detected by EGRET and will lead to a better characterization of the diffuse background.

5.2. *Pulsars*

Detailed observation of known sources and detection of new ones, with excellent time resolution, should help distinguish among the proposed emission models.

5.3. *Supernova Remnants (SNRs)*

For SN candidates, the LAT sensitivity and resolution will allow mapping to separate extended emission from the SNR from possible pulsar components, and thus to measure the possibly differing energy spectra.

5.4. *Unidentified EGRET Sources*

About 170 sources detected by EGRET have not yet been identified with known sources. The increased resolution and sensitivity of GLAST will help us to complete the identification of some of these objects, and generate many new ones.

5.5. *Gamma-Ray Bursts (GRBs)*

EGRET fully detected 4 GRBs. GLAST is expected to see about 200 per year (above 100 MeV) with about a third of them localized to better than ten arcminutes. The LAT and GRB together will measure spectra over seven decades of energy. In addition, the small dead-time of the LAT will allow us to better characterize the short-time-scale GRBs.

5.6. *Active Galactic Nuclei (AGN) and Blazars*

EGRET detected 70-90 AGN. GLAST is expected to detect many more, perhaps 5,000-10,000. In addition, the GLAST energy range is broad, overlapping those of ground-based experiments for good multi-wavelength coverage. The wide field-of-view will allow GLAST to monitor AGN for time variability on many scales. GLAST, in combination with TeV observatories will probe the complex spectra and variability of blazars.

5.7. *Dark Matter Searches*

The possibility of the detection of decays of the neutralino, the lightest super-symmetric particle, depends strongly on its mass. The higher sensitivity and better source resolution will allow us to extend the energy range of the search.

6. Components of the LAT

6.1. *Tracker (TKR)*

The tracker is made of 75 m^2 of silicon wafers, which is the largest area by far ever deployed in a space-faring instrument. Aluminum strips, over 880,000 in all, are embedded in the wafers on a 228-μm pitch. Pairs of planes measuring orthogonal directions alternate with tungsten converter foils, with about 1.5 radiation lengths in the full detector (at normal incidence). The device is self-triggering and has a measured noise occupancy of less than 10^{-6} per strip at an efficiency of greater than 99%. The total power consumed by all 16 tracker modules is about 175 watts.

6.2. *Calorimeter (CAL)*

Each of the sixteen calorimeter modules is composed of eight hodoscopic layers of twelve CsI(Tl) crystals, with dimensions 2x2.7x33 cm^3, 1536 in all, with a total weight of about 1500 kg. Alternate layers are aligned in orthogonal directions. A showering gamma ray sees about 8.5 radiation lengths (for normal incidence). Each crystal is read out from both ends, using the light attenuation in the crystal to provide a measurement of the position along its length. Two PIN photodiodes at each end provide for four overlapping gain ranges. Like the tracker, the calorimeter is self-triggering. The power consumed by the calorimeter electronics is about 90 watts.

6.3. *Anticoincidence Detector (ACD)*

The ACD is a five-sided box which fits over the tracker. It is composed of 89 scintillator tiles, 1 cm thick. Segmentation of the detector minimizes self-vetoing due to backsplash from high energy electromagnetic showers in the calorimeter. Wave-shifting fibers are embedded in the scintillator, and the fibers carry the scintillator light to phototubes situated below the towers. The towers are overlapped where possible to reduce gaps, and the remaining gaps are covered with ribbons of scintillating fibers. The charged particle efficiency is greater than 99.97%. Total power consumed is less than 15 watts.

7. Status of the LAT Construction

As of this writing, all flight tracker and calorimeter modules have been completed. Eight TKR-CAL towers have been assembled into the flight grid and run on cosmic rays, and four more have been installed. The ACD is complete and has been delivered to SLAC for integration. Integration is on schedule for the LAT to be delivered for test and observatory integration in January 2006.

8. LAT Software

There are two faces of LAT software, one looking towards high-energy physics and the other towards astronomy. The first is concerned with reconstructing the parameters of each event, and with separating the gamma-ray signal from the charged-particle, albedo, and other backgrounds. The second is concerned with finding sources, measuring spectra, and producing catalogs. Some of the software components span the two aspects.

8.1. *Event Simulation and Reconstruction*

Much of the software used in the simulation and reconstruction of events in the LAT is home-grown, but we tried to avail ourselves as much as possible of standard freeware for object-oriented I/O, code management, file version management, code documentation, etc. GLEAM, our simulation, reconstruction and analysis program, is written in C^{++} using the Gaudi [8] framework. For simulated data, particles are generated according to various source models and propagated in a model of the LAT, using GEANT4 [9]. The response of the detector is modeled to produce digitized data, which are then reconstructed, both using realistic calibrations (electronic gains, detector misalignments, etc.). Real data is fed into exactly the same software, allowing meaningful comparisons to be easily made. Figure 2 is a display of an actual photon detected during testing of the eight-tower configuration with surface cosmic rays, and reconstructed by GLEAM. Figure 3 shows a large-angle cosmic-ray muon event.

An automated pipeline has been developed which allows us to simulate and/or analyze large numbers of events with minimal manual intervention. All the data taken during cosmic-ray testing of the apparatus has been put through this pipeline, which ultimately will be connected to GLAST's publicly accessible data archive.

8.2. *Science Tools*

The science tools operate on the output of GLEAM processing, namely, lists of photons with origins, energies, and times. The data will be stored in HEASARC [10] in FITS [10] format, and will be examined and manipulated with a series of FTOOLs [10] specific to this mission, but using standard interfaces. In a typical analysis, data will be downloaded from the archive, count and exposure maps will be produced according to the criteria for that analysis, and, for example, sources will be identified and characterized, by using a succession of science tools. The results might be combined with information from other experiments using such high-level tools as Xspec [10], from the Xanadu [10] suite.

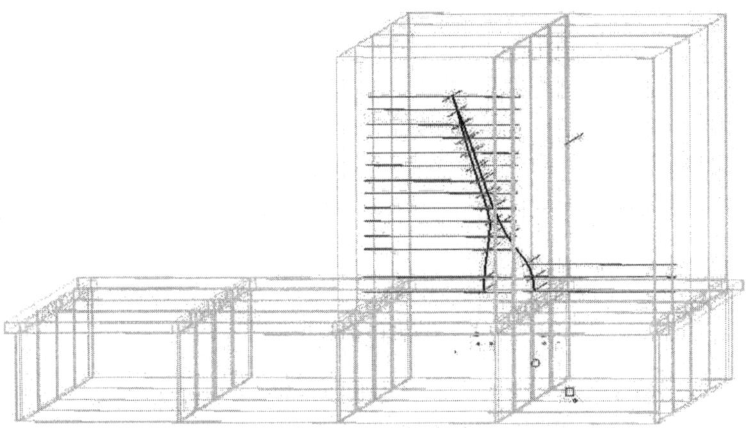

Figure 2. A low-energy gamma ray detected and reconstructed during the commissioning of the 8-tower assembly. The outline of the instrument is shown, as well as the hit strips in the tracker and the centroid of the energy deposits in the calorimeter crystals. The downward-pointing vee is the reconstructed electron-positron pair. The ACD has not yet been incorporated into the instrument.

8.3. *Data Challenges*

An important strategy for preparing the software for flight is a series of data challenges. These consist of end-to-end trials of the software, from

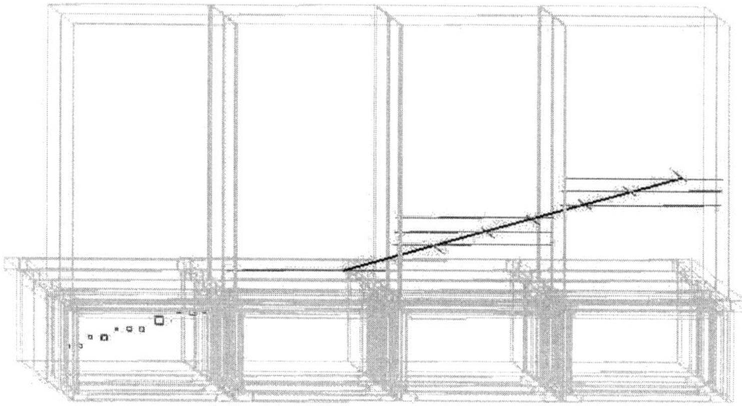

Figure 3. A cosmic-ray muon detected and reconstructed in the 8-tower assembly. This event demonstrates the large field-of-view of the LAT.

simulation to science results. They are designed to provide real deadlines for code projects, to help us learn what works and what needs improvement, to foster teamwork and communication among developers, and to involve new users across the collaboration. There are three progressively more elaborate data challenges in the series: one has already taken place, the second is in the planning stage.

The first, DC1, had modest goals: we produced one day of simulated all-sky survey data with no background, which contained some simple physics and a few surprises. The second, DC2, will take place in early 2006. It will include one month of simulated data (including the South-Atlantic Anomaly), a rich background, variable sources, and integration of data from the GBM. The instrument response will include effects such as dead channels and finite buffer sizes. DC3, in 2007, will be a simulation of full flight-science production.

Soon after DC3, we will embark on the real data challenge: the launch of GLAST is scheduled for August 2007, which will mark the beginning of a fruitful five- to ten-year mission.

9. Acknowledgments

The author is pleased to acknowledge the help of the entire GLAST LAT team, and particularly of his colleagues in the Science Analysis Software Group. This work was supported in part by the U.S. Department of Energy under contract number DE-AC02-76SF00515.

References

1. Online at http://glast.gsfc.nasa.gov.
2. Online at http://gammaray.msfc.nasa.gov/gbm.
3. Online at http://glast.stanford.edu.
4. Kanbach, G., et al., Space Science Reviews 49, 69 (1988).
5. Thompson, D. J., et al. Astrophysical Journal Supplement 86, 629 (1993).
6. Gehrels, N., Chipman E., and Kniffen, D. A., "The Compton Observatory in Perspective," Compton Gamma-Ray Observatory, AIP Conference Proceedings, 290, ed. N. Gehrels, et al., 3 (1993), AIP, New York.
7. Online at http://cossc.gsfc.nasa.gov.
8. Online at http://wwwasd.web.cern.ch/wwwasd/geant4/geant4.html.
9. Online at http://lhcb-comp.web.cern.ch/lhcb-comp/Frameworks/Gaudi/.
10. Online at http://heasarc.gsfc.nasa.gov/docs/heasarc/.

STATUS OF THE ARGO-YBJ EXPERIMENT

P. CAMARRI
FOR THE ARGO-YBJ COLLABORATION
University of Roma "Tor Vergata" and INFN Roma 2
Via della Ricerca Scientifica, 1,
00133 Roma, Italia
E-mail: Paolo.Camarri@roma2.infn.it

The ARGO-YBJ experiment is designed for EAS detection in the primary energy range 100 GeV - 10 TeV. Its full-coverage feature allows to bring down the lower energy threshold about one order of magnitude with respect to sampling EAS detectors. Here the main features of the detector are recalled and the present status of the experiment is described.

1. Introduction

The need for ground-based detectors for cosmic-ray astrophysics and gamma-ray astronomy comes from the limitations of satellite measurements at energies higher than $\sim$ 100 GeV. Due to the decreasing power law for gamma-ray flux, it would take years to collect significant statistics from a high-energy gamma source on a $\sim 1\ m^2$ detector at those energies. Ground-based detectors can investigate a much wider energy range by exploiting the EAS amplification effect. The two main experimental techniques exploited so far for EAS detection on the ground are based on air Cherenkov telescopes (ACTs) and sampling EAS arrays.

ACTs have been usually used for thorough investigation of already known gamma-ray sources, due to their very low energy threshold ($<$ 100 GeV), high background rejection, high sensitivity and good energy resolution. The discovery power of typical ACTs is strongly limited by their low duty cycle ($<$ 10%) and small field of view ($\Delta\theta$ less than few degrees).

EAS arrays at sea level, on the other hand, have a much higher energy threshold ($>$ 10 TeV), moderate background rejection, modest sensitivity and energy resolution. However, they have a very high duty cycle and a large field of view, providing a huge potentiality for the discovery of new sources.

In order to improve the performance of ground-based EAS arrays, two main goals must be pursued:

- lowering the energy threshold (< 500 GeV);
- increasing the sensitivity to $\sim 1/10$ the Crab gamma-ray flux.

A possible strategy in this respect may be summarized as follows:

- high-altitude experimental site;
- secondary photon conversion;
- increased sampling, approaching 100%.

The ARGO-YBJ detector [1] is designed to meet the requirements mentioned above. The main physics goals of the experiment are the following:

- γ-ray astronomy: search for galactic and extra-galactic sources at energy greater than a few hundred GeV;
- diffuse γ-rays from the galactic plane and SNRs;
- physics of γ-ray bursts;
- cosmic-ray physics, particularly the spectrum and composition around the knee (E $>$ 10 TeV);
- Sun and heliosphere physics.

Here the main features of the detector are summarized, and the early experimental measurements with a reduced setup are reported.

2. The experimental setup

The ARGO-YBJ experiment is located at Yang Ba Jing (Tibet, China), $90^\circ\,31'\,50''$ E, $30^\circ\,06'\,38''$ N, 4300 m above sea level. The detector is a full-coverage layer of Resistive Plate Chambers (RPCs) [2]. A scheme of the detector section is shown in Fig. 1. The sensitive part consists of a 2 mm thick gas gap enclosed between two high-resistivity plastic-laminate plates ($\sim 5 \times 10^{11}\,\Omega\,$cm volume resistivity). The plates are kept at a fixed distance by a grid of cylindrical PET spacers (1 cm diameter, 10 cm pitch), and the detector is operated with a three-component gas mixture: Ar/i-$C_4H_{10}/C_2H_2F_4$ = 15/10/75. The outer faces of the resistive plates are painted with a thin graphite layer, to apply a uniform voltage to the whole gas gap (7.2 kV at the experimental site in Tibet). The discharges produced by the passage of charged particles through the gap are picked up by a read-out strip plane (7 cm strip pitch), and transmitted to the front-end electronics (directly soldered on both sides of the strip plane) for digital

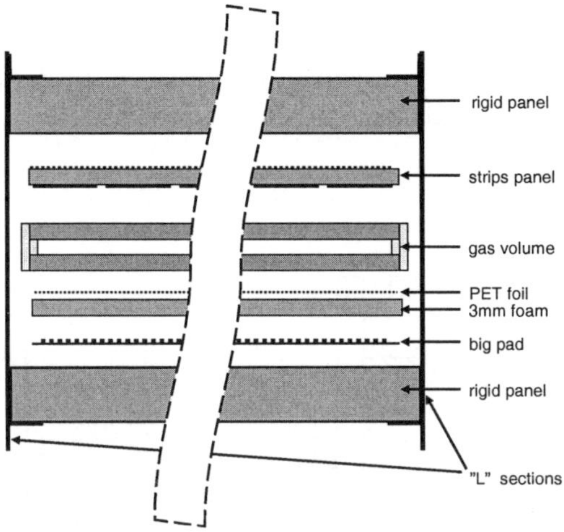

Figure 1. Schematic cross-view of one ARGO-YBJ chamber.

processing. The intrinsic RPC time resolution at the chosen operating voltage is about 1 ns. The space resolution is provided by the area of a single strip (6.5×62 cm^2). One RPC chamber is assembled in the shape of a box of $2850 \times 1225 \times 47$ mm^3. The gas volume is kept in a fixed position by two external panels (200 μm thick foil glued on a 1.5 cm foam layer. The read-out strip plane is made of two 17 μm thick copper foils glued on a 3 mm thick foam layer; the copper foil facing the gas volume is cut into strips of size 7.0×62.0 cm ^{2}. 8 contiguous strips make a logical "pad", and the OR of the 8 strips gives the time signal for the TDCs. The front-end electronics is based on a GaAs custom chip, consisting of 8 discriminator channels each providing a single-ended ECL level at the output. The power consumption is about 25 mW per channel. A special Zener-diode protection is applied at the input stage of each front-end channel, providing a safe operation up to about 8 kV, necessary in case of high-density events ($\approx 10^3$ particles per strip).

On the opposite side of the gas volume with respect to the read-out strip plane, each chamber is instrumented with two adjacent analog charge readout planes (17 μm thick copper plane glued on a 190 μm thick PET foil), called "big pads". These are used to measure the induced analog signal from the two halves of a RPC, and provide a complementary measurement with respect to the digital information from the strip plane.

The ARGO-YBJ chambers are placed one next to the other in a full-coverage array, surrounded by a "guard ring" to improve the detection of showers not fully contained in the central area. The complete layout is shown in Fig. 2, where two subsequent zooms show the detailed structure of a *cluster* (a group of 12 adjacent chambers) and the pads inside a chamber.

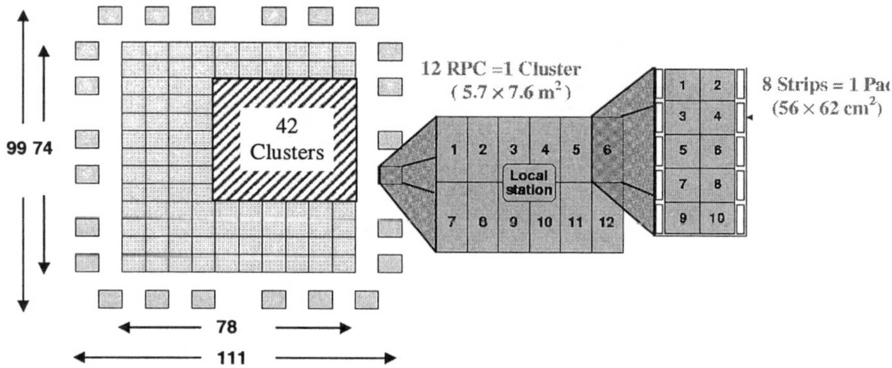

Figure 2. Layout of the ARGO-YBJ experiment. The central full-coverage area is surrounded by a "guard ring". The 42 clusters used for the preliminary measurements are evidenced. The side lengths of the experimental setup are in meters. Zooms show the structure of a cluster (12 adjacent chambers) and the pad positions inside a chamber.

The trigger for the events is provided by a minimum required pad multiplicity in the central part of the detector (60 in the preliminary measurements with 48 clusters), with space-time consistency for a shower front. Beside the main trigger, a measurement of the pad rate from each cluster (with a 0.5 s integration time) was introduced to detect unexpected changes in the cosmic-ray flux, which is a feature of GRBs and solar flares, for instance.

The Detector Control System (DCS) is crucial includes all the control and monitoring tasks necessary to guarantee a safe and reliable operation of the detector. The main task of the DCS are: high-voltage control and monitoring, gas-flow monitoring, monitoring of the environmental parameters (indoor and outdoor temperature, barometric pressure), current monitoring and count-rate monitoring for each chamber for detailed diagnostics. Its role is crucial to protect the detector in case of anomalous behaviour of any of its parts.

3. Preliminary measurements

From January till October 2004, data have been taken with a reduced setup (16 clusters, about 700 m^2 sensitive area) in order to optimize the operational parameters of the detector, to monitor its long-term performance and to test the off-line reconstruction algorithms. Starting from the end of 2004, the experimental setup was extended to 42 clusters (about 1900 m^2 sensitive area).

In Fig. 3 an example of a shower reconstruction is shown. The upper plot represents the space-time behaviour of the shower front, while the lower plot shows the projection of the hit pads on the detector plane. This shower is entirely contained within the detector area.

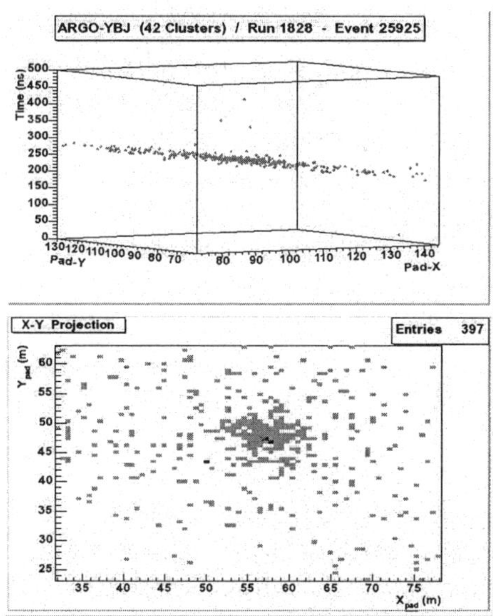

Figure 3. Space-time profile and hit projection on the detector plane for one fully-contained reconstructed shower on 48 clusters.

Another reconstruction is shown in Fig. 4. This is an example of a shower extending beyond the detector limits, although the core is still contained within.

In Fig. 5 the preliminary reconstructed zenith distribution for the events collected on 42 clusters is shown [3].

A preliminary measurement of the analog charge distribution in the

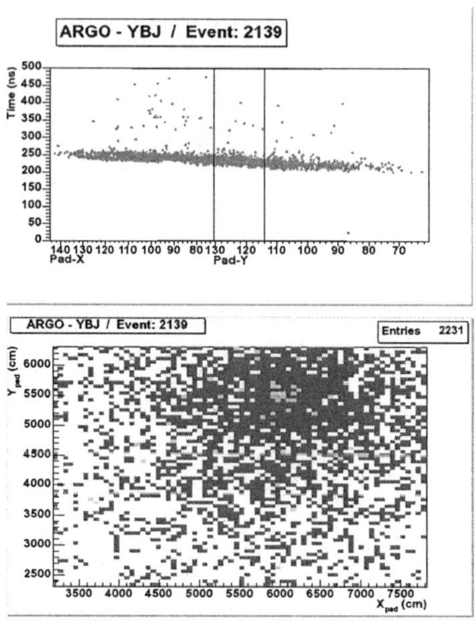

Figure 4. Space-time profile and hit projection on the detector plane for one reconstructed shower extending beyond the detector limits.

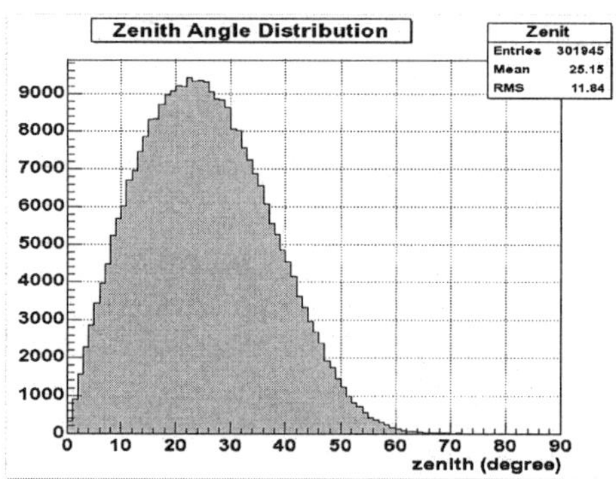

Figure 5. Preliminary zenith-angle distribution for the reconstructed events on 48 clusters.

events was made on 4 clusters (about 180 m^2) to test the system capabilities. In Fig. 6 the charge distribution for a fully contained event is shown: the core position is clearly visible.

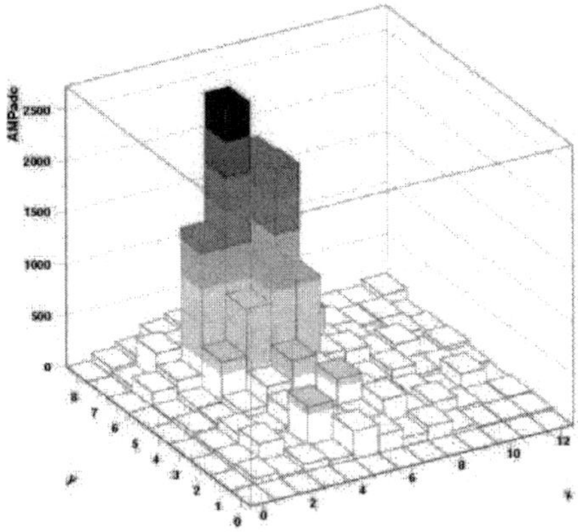

Figure 6. Analog charge distribution on the big pads of 4 clusters for a fully-contained event.

4. Conclusions

During the measurement session with a reduced set-up, the performance of the ARGO-YBJ detector was in good agreement with the expectations. Preliminary results based on the collected data will be presented at the International Cosmic Ray Conference in Pune, India (August 3-10, 2005). The completion of the experimental setup is foreseen in 2006.

References

1. The ARGO-YBJ Collaboration. "Astroparticle physics with ARGO", Technical Proposal (1996); "The ARGO-YBJ Project", Addendum to the Proposal (1998). http : //argo.na.infn.it/argo_pub_artic.html.
2. R. Santonico and R. Cardarelli; Nucl. Instr. and Meth. A 377 (1981), 187. R. Cardarelli and R. Santonico; Nucl. Instr. and Meth. A 263 (1988), 200.
3. K. Calabrese Melcarne; contribution to this Workshop.

GAMMA AIR WATCH (GAW) - AN IMAGING ATMOSPHERIC CHERENKOV TELESCOPE LARGE WITH LARGE FIELD OF VIEW

T. MINEO, G. CUSUMANO, G. AGNETTA, B. BIONDO, O. CATALANO, F. CELI, S. GIARRUSSO, A. LA BARBERA, G. LA ROSA, M.C. MACCARONE, A. MANGANO, F. RUSSO, B. SACCO, A. SEGRETO

IASF-Pa/INAF, via U. La Malfa 153, 90146, Palermo, Italy
E-mail: mineo@pa.iasf.cnr.it

G. DI COCCO

IASF-Bo/INAF, via Gobetti 101, 40129 Bologna, Italy

P. ASSIS, P. BROGUEIRA, J. COSTA, M.C. ESPIRITO SANTO, M. PIMENTA, B. TOMÈ

LIP, Av. Elias Garcia 14, 1o, 1000-149 Lisboa, Portugal

C. DELGADO

IAC, Via Lactea s/n, E-38200, La Laguna, Tenerife, Spain

P. GALEOTTI, C. VIGORITO

Dept. of Physics, University of Turin, Via Giuria 1, 10125, Torino, Italy

M. MOLES, F. PRADA, M.A. SANCHEZ, A. DE UGARTE POSTIGO

IAA, Camino Bajo de Huétor 50, 18080 Granada, Spain

P. VALLANIA

IFSI-To/INAF, Cs Fiume 4, 10133 Torino, Italy

GAW, an acronym for Gamma Air Watch, is an experiment to test the feasibility of a new generation of imaging atmospheric Cherenkov telescopes with large field of view capability. GAW is an array of three identical telescopes disposed at the vertexes of an equilateral triangle, ~100 m side. Two main features characterize GAW with respect to all the existing and presently planned ground-based Cherenkov telescopes. The first difference concerns the optics system: GAW uses a Fresnel refractive lens (Ø2.13 m) as light collector instead of classical reflective mirror. The second main difference is that the detector at the focal surface operates in single photoelectron counting mode instead of the usual charge integration one. The GAW array is planned to be located at the Calar Alto Observatory site,s Spain, 2150 m a.s.l. During its first phase, only 6° × 6° of the focal plane detector will be implemented; in a second phase, the focal plane will be enlarged to cover a field of view of ± 24°. GAW is a collaboration effort of Research Institutes in Italy, Portugal and Spain.

1. Introduction

Despite its youngness, Very High Energy (VHE) gamma-ray astronomy is considered a legitimate astronomical discipline with established sources, steady and variable, galactic and extragalactic, that provide deep implications in the theoretical models. Presently, the main objectives of the ground-based VHE gamma-ray experiments are; the improvement of the flux sensitivity above 100 GeV to increase the number of VHE gamma-ray sources and give more details on their properties; the extension of the observation range towards lower energies (few tenths of GeV) so to meet and overlap the energy window proper of the gamma-ray satellite experiments. The first objective is achieved using arrays of telescopes as VERITAS (Wekees at al. 2002), HESS (Bernlöhret al. 2003, Vincent et al. 2003), or CANGAROO III (Kabuki et al. 2003, Enomoto et al. 2002), which, thanks to their stereoscopic observational approach, furtherly increase the sensitivity capability of IACT telescopes. A lower energy range is being exploited with arrays of solar heliostats as STACEE (Chantell et al. 1998), Solar-2 (Tümer et al. 1999), CELESTE (Quebert et al. 1998), or GRAAL (Arqueros et al. 1999); moreover, the telescope MAGIC (Baixeras 2004), with its 17 m aperture, is going to achieve a similar low energy threshold using the Imaging Atmospheric Cherenkov Technique (IACT).

Nevertheless, astronomical events can occur at unknown locations and/or random in time, and a large field of view (FoV) is then needed to increase their detection probability. To collect light, the present IACT telescopes use large mirror reflectors characterized by FoV of the order of few degrees; they cannot reach larger FoV due to the mirror optical aberrations, rapidly increasing with off-axis angles, and to the shadow of a larger detector onto the reflector. A large FoV is also a basic requirement to perform sensitive survey of the Galactic Plane as well as an estimation of the celestial gamma-ray diffuse emission. Such a survey cannot be easily performed by the ground-based IACT telescopes: as a result of their reduced FoV, such telescopes can survey the sky only if long exposure times are considered.

To overcome such limitations, an alternative solution could come from the use of refractive optics, as the Fresnel lenses: they can achieve large FoV with imaging capability suitable to the quite coarse structure of the Cherenkov image; they maintain imaging stability against deformation; there is no central obstruction of the detector at the focal surface.

GAW, acronym for Gamma Air Watch, is an experiment, currently under definition, to test the feasibility of a new generation of IACT telescopes that join high flux sensitivity with large field of view capability using Fresnel lens, stereoscopic observational approach, and single photoelectron counting mode.

2. The experiment

GAW is conceived as an array of three identical IACT telescopes disposed at the vertexes of an equilateral triangle. A detailed description of GAW is given in the GAW proposal*; here we report two specific main features of the array as R&D experiment for a new generation Cherenkov telescopes.

A refractive optical system characterizes GAW: its light collector is a non commercial Fresnel lens (Ø2.13 m) with focal length of 2.55 m and standard thickness of 3.2 mm. The lens material is UltraViolet (UV) transmitting acrylic with a nominal transmittance of 95% from 330 nm to the near InfraRed; this material joins high transmittance and small refraction index derivative at low wavelength, reducing chromatic aberration effects. The lens design is optimized to have, at the wavelength of maximum intensity of the Cherenkov light convolved with the detector response (~360 nm) a quite uniform spatial resolution up to 30° (full angle) suitable to the requirements of the Cherenkov imaging. The baseline optics module for the GAW prototype is a single-sided, flat Fresnel lens optimized for a ±12° field of view. The flat lens is composed by a central core (Ø50.8 cm) surrounded by a corona of petals extending for 40.6 cm out from the radius of the core, and a second outer corona of petals extending for 40.6 cm more. The central

*in preparation

core will be made with constant depth aspheric grooves; the petals will have constant width aspheric grooves. The assembled lens will be constituted by one single piece for the central core, 12 pieces for the first petals ring, and 20 pieces for the outest petals corona. A spider support will maintain all the pieces together. The project of the optical system is a joint effort of the IASF/INAF Institute in Palermo and of the Fresnel Technologies, Fort Worth, Texas, which will manufacture the lens.

The second main feature characterizing GAW is the detector working mode used. The focal surface detector of each telescope consists of a grid of MultiAnode PhotoMultiplier Tubes (MAPMT) manufactured by Hamamatsu, series R7600-03-M64; the number of active channels (order of 104) forming the detector at the focal surface makes it basically a large UV sensitive digital camera with high resolution imaging capability. The large dead area of the chosen MAPMT induces a low geometrical efficiency factor of ~50% on photon detection; in order to correct that, each MAPMT pixel is coupled to a Light Guide (LG) which allows to uniformly cover the FoV with a 80% average absorption (due to the LG). The specific feature is that the GAW electronics design is based on single photoelectron counting mode (front-end), instead of the charge integration method widely used in the IACT telescopes, and on free-running method (data taking and readout). The single photoelectron counting mode method is a well-established technique and it is used to measure the number of output pulses from the photosensors corresponding to incident photons. Small pixel size is required to minimize the probability of photoelectrons pile-up within intervals shorter than the given sampling time of 10 ns (GTU, Gate Time Unit). In such working mode, the electronics noise and the MAPMT gain differences are kept negligible allowing lowering the energy threshold in spite of the relatively small dimension of the GAW light collector. The free-running method makes use of cyclic memories to continuously store system and ancillary data at a predetermined sampling rate; once a specialized trigger stops the sampling procedure, data are recovered from the memories and ready to be transferred to a mass memory.

The single photoelectron counting mode, together with the stereoscopic observational approach, will guarantee an energy threshold of the order of few hundreds of GeV in spite of the relatively small dimension of the lens.

GAW is planned to be located at the Calar Alto Observatory (Sierra de Los Filabres, Almeria, Spain, 2150 m a.s.l.). Two phases are foreseen for the project:

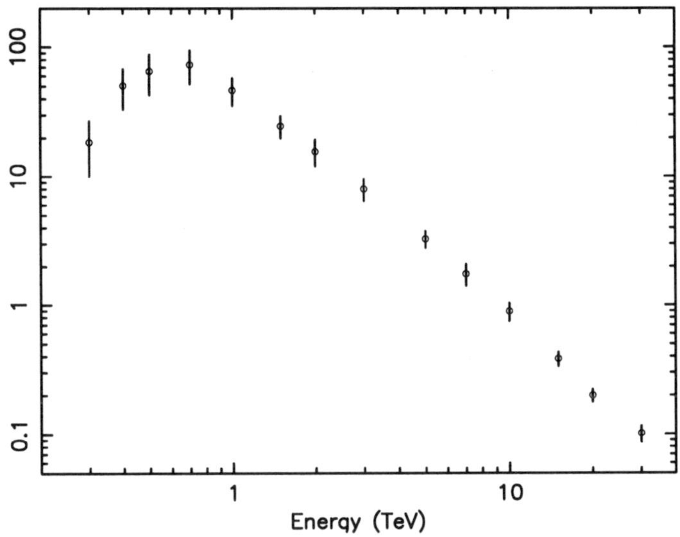

Figure 1. GAW differential detection rate of the Crab Nebula vs energy.

- Firstly, only part of the GAW focal detector will be implemented to cover a FoV of $6° \times 6°$; the detector will be mounted on a rack frame and shifted along the full size of the focal surface. Under this configuration the sensitivity of the telescopes will be tested observing the Crab Nebula with on-axis and off-axis pointings up to 20°. GAW will also monitor the VHE activity of some flaring Blazars.
- In a second phase, once the feasibility of the method proposed has been proved, the focal plane detector will be enlarged to cover a FoV of $\pm$ 24°. We plan to survey a region of $360° \times 60°$ of sky pointing along different North-South directions.

3. GAW performance

To evaluate the GAW expected performance, a complete end-to-end simulation chain has been developed, starting from the physical process to the event reconstruction and analysis, through the effect of the atmospheric absorption and the detector response. The first step of the simulation chain, performed by using the CORSIKA code[†], mainly concerns the generation of Cherenkov light, at level of single photons, associated to air showers induced

[†]http://www-ik.fzk.de/corsika

by different primaries; moreover, it includes the effects of the atmospheric absorption and a set of detector parameters (optics transmission, MAPMT photoelectron efficiency, ...) values of which are considered nominal at this stage of the project (GAW in reduced configuration). This step is computationally time-consuming and a library of more than 6000 simulated air showers (protons and gamma) have been produced in the energy range 0.3-30 TeV as well as Crab-like spectrum and mono-energetic primaries, on-axis and off-axis.

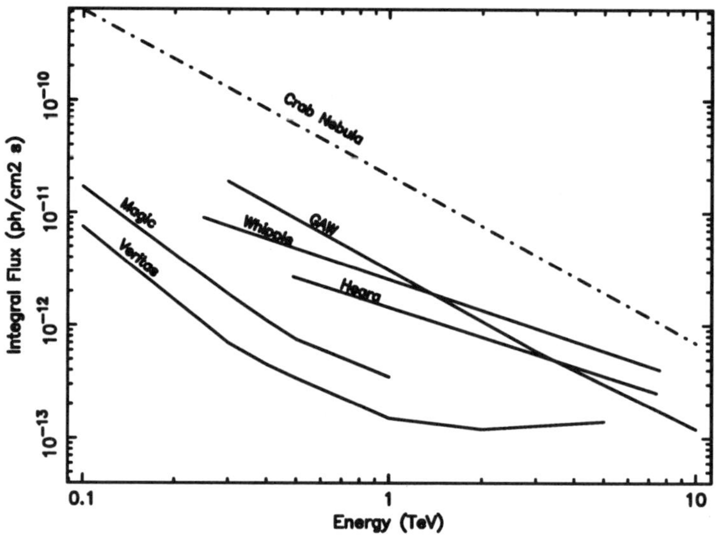

Figure 2. GAW sensitivity (5σ) detection for Crab-like sources reachable during the first phase ($6° \times 6°$ FoV) in 50 hours observations.

The Cherenkov files are then analyzed by a proper user-defined home-made code that includes other peculiarities of the detector which can be easily updated whenever the hardware status of the telescope is modified. Among these, the GAW geometrical configuration, the optics size and its spread function, the effect of the light guides, the remaining efficiency factors, as well as the trigger electronics and the average expected value of light diffuse background are included. A threshold of 14 photoelectrons per event in each telescope, together with the coincidence in the other two telescopes is applied; this choice reduces the fake trigger rate due to the diffuse background to a negligible level. A first group of simulation has been generated considering the three telescopes located at the vertices of a triangle 100 m side. As in the standard analysis of stereoscopic observations, the reconstruction of the arrival direction is obtained superimposing the Cherenkov images of the three telescopes in a common plane and determining the point of minimal distance among the major axes of the image ellipses. The direction of each major axis is determined by minimizing its distance with respect to the photoelectrons position after suitable suppression of the more external points, mainly due to the diffuse light background.

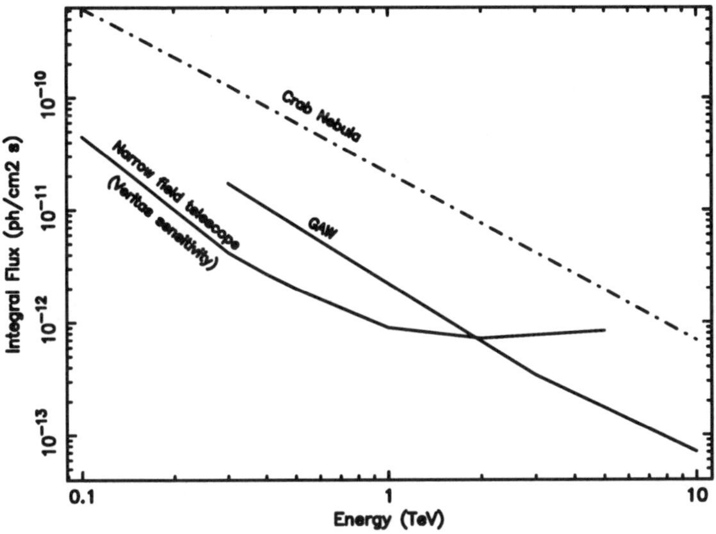

Figure 3. GAW capability of GAW to perform deep sky survey with enlarged FoV (24° x 24°) in two years observation

Fig. 1 shows the differential detection rate of the Crab Nebula as expected to be observed by GAW. We define the detection rate as the rate of gamma-ray events remaining after the cosmic-ray background rejection. The energy threshold of GAW will be about 300 GeV with a peak at about 700 GeV. The performance of GAW is summarized by its integrated flux sensitivity as function of the energy. Figure 2 shows the minimum integrated flux for detection of a 5 σ excess, with at least more than 10 photons, in 50 hours of observation of a gamma-ray source with a Crab-like spectrum (dN/dE $\propto E^{-2.5}$) evaluated during the first phase ($6^\circ \times 6^\circ$ FoV). Figure 3 shows the capability of GAW to perform deep sky survey with enlarged FoV ($24^\circ \times 24^\circ$).

4. Conclusion

GAW is and R&D and path-finder experiment to test the possibility of a new generation of IACT telescopes that join high flux sensitivity with large field of view capability using Fresnel lens, single photon counting mode, and stereoscopic observational approach. Thanks to the single photon counting mode, which allows to operate with a very low photoelectrons threshold ($\sim$1/8 respect to the value required by the present IACTs), an enlargement of the GAW optics diameter could allow a very deep sensitivity observing a field of view one hundred times larger than what the present IACT telescopes can reach.

References

1. F. Arqueros, in *XXX IMFP (Jaca) 2002. Published in Nucl. Phys. B* **114**, 253 (2003).
2. C. Baixeras, *Nucl.Phys.B (Proc.Suppl.)* **114**, 247 (2003).
3. K. Bernlöhr, O. Carrol, R.Cornils, et al., *Astroparticle Physics* **20**, 111 (2003).
4. M.C. Chantell, D. Bhattacharya, C.E.Covault, et al., *Nucl. Instr. Meth. A* **408**, 468 (1998).
5. R. Enomoto, T. Tanomori, T. Naito, et al., *Astroparticle Physics* **16**, 235 (2002).
6. S. Kabuki, K. Tsuchiya, K.Okumura, et al., *Nucl. Instr. Meth. A* **500**, 318 (2003).
7. J. Quebert, et al., in *Proc of 'Towards a Major Atmospheric Cherenkov Detector - IV' (Padova, Italy), ed. M.Cresti)* 248 (1995).
8. T. Tümer, D. Bhattacharya, U. Mohideen, et al., *Astroparticle Physics* **11**, 271 (1999).
9. P. Vincent, J.P. Denance, J.F. Huppert, et al., in *Proc. 28th ICRC, Tsukuba (University Accademy Press, Tokio)*, 2887 (2003).
10. T.C. Weekes, H. Badran, S.D. Biller et al., *Astroparticle Physics* **17**, 221 (2002).

Part II

Topics in Fundamental Physics

FRONTIERS OF HIGH ENERGY COSMIC RAYS

MÁRIO PIMENTA

LIP, Av.Elias Garcia, 14-12, 1000-149 Lisboa, Portugal
IST, Av. Rovisco Pais,1049-001 Lisboa, Portugal

High energy cosmic rays are a window to the Universe, an opportunity to Particle Physics but they represent an enormous challenge to build new detectors.

The energy spectrum of cosmic rays extends over more than 11 decades of energy, following an almost perfect power law (see figure 1) [1]. The fluxes are quite high at low energies (up to 1 $particle/m^2/s$ for energies of the order of the GeV), but extremely small at the highest energies (1 $particle/km^2/century$ at 10^{20} eV). At the far end of the spectrum (figure 2), just a few events have been observed, mainly in two experiments: AGASA [2] and HiRes [3]. While AGASA seems to be compatible with no slope change in the spectrum at these extreme energies, HiRes seems to suggest a dramatic drop. The statistical significance of the difference between the two experiments is however slightly above 2σ [4]!

At energies above 10^{19} the deflexion by the galactic magnetic field is small and so charged very high energy cosmic rays are indeed an astronomical channel. The geographical distribution of the highest energy data (above 4 10^{19} eV) collected by AGASA is isotropic at large scales but shows some clustering at smaller scales: one triplet and six doublets with $\Delta(\Theta) = 2.5^0$ [5].

There are many papers discussing possible acceleration mechanisms able to produce particles with such energies. The regions where such events may occur can be characterised by two variables: the size of the region and the intensity of the existing magnetic fields. It is possible to have confinement either in small regions, if there is an intense magnetic field, or, if the magnetic field is weak, in extended regions. Such a feature is usually summarised in the so called Hillas plot (figure 3) [6]. There are several possible sources for 10^{20} eV UHECR but at higher energies it is much more difficult to find good candidates. Furthermore, energy losses due to

Bremsstrahlung have to be taken into account whenever the accelerations are very large [7].

In any case, the observed UHECR would have to propagate through very large distances before reaching the Earth. But there is no empty space. The space is, as it is well known, filled with the Cosmic Microwave Background (CMB). The interaction of the UHECR with the CMB γs can be quite dramatic if the centre-of-mass energy is above the first inelastic channel threshold. The net result is that UHECR with energies at the source well above 10^{20}eV would have dropped to about 10^{20}eV by the time they reach Earth. This is the famous GZK effect [8]. GZK is not however a real cut-off. The threshold energy depends on many parameters (the type of the sources and its distribution in space, the existing magnetic fields, the particle type, ...).

But if AGASA is right how to avoid the GZK? One possible solution is to imagine that UHECR are particles with small electromagnetic couplings but able to originate normal air showers. Possible candidates may be neutrinos. In the Z burst model [9], very energetic neutrinos collide with the cosmological neutrino background producing energetic Z^0 bosons near our galaxy. It is an elegant model, but it implies the existence of intense and monochromatic neutrino beams with an energy around 4 10^{21}eV! There are many other proposed solutions [10] such as: enhanced neutrino-nucleon cross-sections, the existence of "new" particles (axions, glueballinos, ...) or even the violation of Lorentz symmetry which could lead to a change on the relevant thresholds. However in this kind of models (the bottom-up scenarios), UHECR would be produced as secondaries of very energetic "ordinary" particles ($E > 10^{22}$eV). There is another way out (the top-down scenarios): UHECR may be decay products of super-heavy particles, which would be produced continuously by cosmological "objects" left over after the phase transitions that the Universe has suffered, or even produced directly in the early Universe. The challenge is to increase the statistics of observed UHECR. If this is achieved, models can be disentangled looking at the energy spectrum power law, the UHECR composition, the geographical distribution, the possible counter-parts in other wavelengths - all a program!

Another major challenge in cosmic ray is the observation of high energy cosmic neutrinos. In fact astrophysical neutrino beams should exist [11]! A safe prediction is the existence of the cosmological neutrinos produced in the interaction of the primordial UHECR with the γ_{CMB} or with clouds of intergalactic gas. These interactions produce collimated pions and kaons, which will originate energetic neutrino beams. But there are many other models: from the Z burst mechanism, to the possible Zevatron sources (AGNs, GRBs, supernovae, ...) and to the top-down scenarios discussed just above (see figure 4). These neutrinos can propagate to Earth almost unscattered and so their detection would open an important channel to observe the Universe. However, so far, there is no experimental measurement above 10^6 GeV. The existing limits, as well as the expected sensitivities of future experiments, are shown in figure 5 .

UHECR are the most energetic beams ever accessible. The centre-of-mass energy of a $10^{19} - 10^{20}$ eV UHECR collision in the atmosphere is around 100-400 TeV, which is well above the present and future man made accelerators. But UHECR fluxes are small, the accessible kinematic region is very narrow (very forward region) and the detection capability is much poorer than in conventional Particle Physics experiments. The solution for these low fluxes is to have huge detection areas. The surface of the Pierre Auger observatory is around 3000 Km^2 [12], while the planned space based experiments like EUSO [13] will cover a surface of around 200000 Km^2. The limit is the Earth surface, which is still a factor 1000 larger.

UHECR interact with the Earth atmosphere originating Extended Air Showers (EAS) composed by billions of particles (mainly electrons and photons). If the primary is a proton or a nuclei (the most probable scenario) then hadronic interactions plays a determinant role in the first stages of the EAS development. However, there is no computable theory for soft hadronic interactions and one must deal with phenomenological models and the corresponding systematic uncertainties. At such energies new effects, as percolation [14], can occur.

Nevertheless, important particle physics results were already obtained or are expected in the near future. For instances: pp cross-sections measurements were extended to higher centre-of-mass energies [15]; heavy quark production is very abundant at these energies and may represent an important channel for QCD studies; new particles (Higgs, SUSY, excited fermions, leptoquarks...) with masses above the TeV can be produced; Lorentz symmetry can be tested at very high boost factors; the existence of extra-dimensions may imply important enhancements in the neutrino-air cross-section [16]. The search for new physics is, therefore, a relevant issue in the next generation of UHECR experiments and an effort for finding new signatures must be pursued. An example of such possible signatures is the so-called double-bang signature. An example would be the observation, in an experiment like EUSO, of the production and the instantaneous decay of a microscopic black hole (first bang), followed, at a measurable distance, by the decay of an energetic tau lepton (second bang) [17].

In the next decade a spectacular increase in the number of observed UHECR is expected (one to two orders of magnitude). Auger is right around the corner. It combines the technique of sampling the shower particles reaching the Earth surface with the technique of detecting the fluorescence light produced by the excitation of Nitrogen molecules by the low energy electrons of the shower. At present, 40% of the 1600 sampling detectors (water tanks) and two of the three sites for fluorescence light detection are already fully working! EUSO, an ESA mission planned for the first years of the next decade, is basically a telescope looking downwards to the Earth in order to collect a fraction of the fluorescence light produced in the interaction of the UHECR with the atmosphere. And there are already plans to go further, OWL [18] is a NASA mission under study which will use two satellites, reaching a field of view about 5 times larger than EUSO. Finally, there are several experiments on the way to the detect the coherent radio emission by very energetic showers[19].

Aknowledgements

I would like to thank Catarina Espírito Santo and Pedro Assis for reading the manuscript and help in preparing these proceedings, and Alessandro De Angelis for the nice atmosphere during all the conferenc

References

1. For a general reference see for example: A.Watson and M.Nagano, Rev. Mod. Phys., 72, 689 (2000).
2. M.Takeda et al., Astro. Part. Phys., 19, (2003) 499
3. D.Bird et al., Astrophys. J., 511, (1999), 739
4. D.De Marco, P.Blasi, A.V.Olinto, Astropart. Phys. 20 (2003) 53
5. N.Hayashida et al., astro-ph/0008102 (2000)
6. A.M.Hillas, Ann. Rev. Astron. Astrophys. 22 (1984) 425.
7. M.V.Medvedv, Physical Review E 67, 045401(R) (2003)
8. K.Greisen. Phys. Rev. Lett., 16, (1996), 748; G.T.Zatsepin and V.A.Kuzmin, Sov. Phys. JETP Lett. (engl. transl.), 4, (1996), 78
9. T.J.Weiler, Phys. rev. Letters 40 (1982) 234; D.Fargion et al., Astrophys. J. 517 (1999) 725
10. G.Sigl, astro-ph/0210049
11. D.V.Semikoz, Theoretical Predictions of Ultra-High Energy Neutrino Fluxes 28th ICRC Tsukuba, 1439 (2003)
12. http://www.auger.org/

13. http://www.euso-mission.org/cgi-bin/pub.cgi?href=welcome.html
14. hep-ph/0507227, J.Dias de Deus et al.
15. M.M.Bock et al., astro-ph/0004232 (2000)
16. M.Cavagla, Int. J. Mod. Phys. A18,1843 (2003)
17. V.Cardoso,hep-ph/0405056
18. http://owl.gsfc.nasa.gov/
19. H.Falcke, astro-ph/0409229

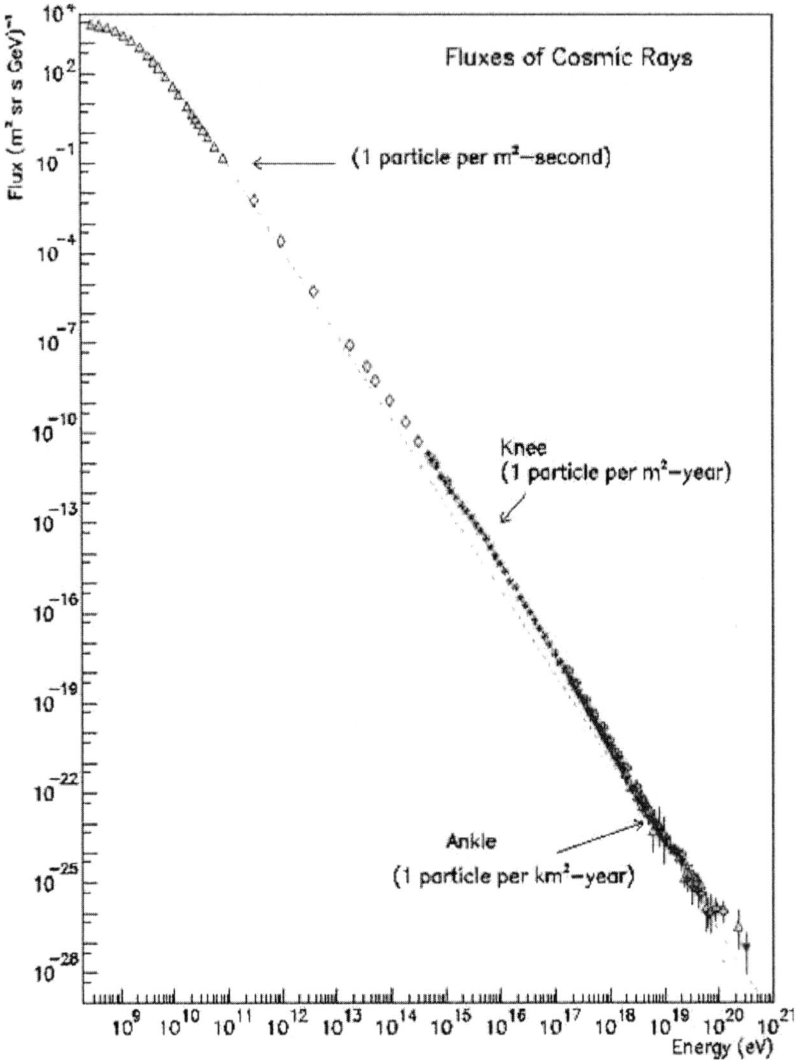

Figure 1. The cosmic ray spectrum .

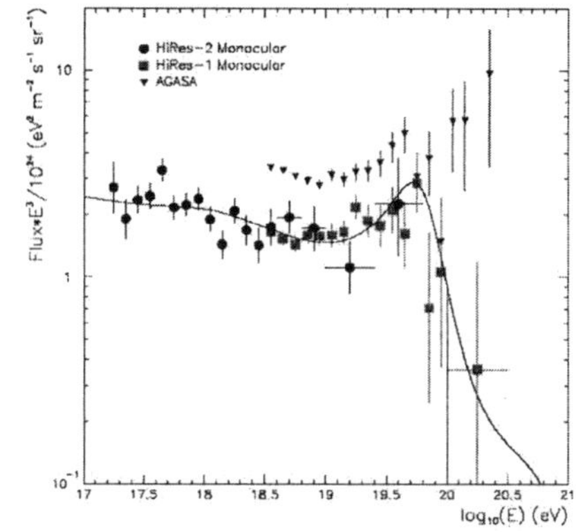

Figure 2. The far end of the cosmic ray spectrum.

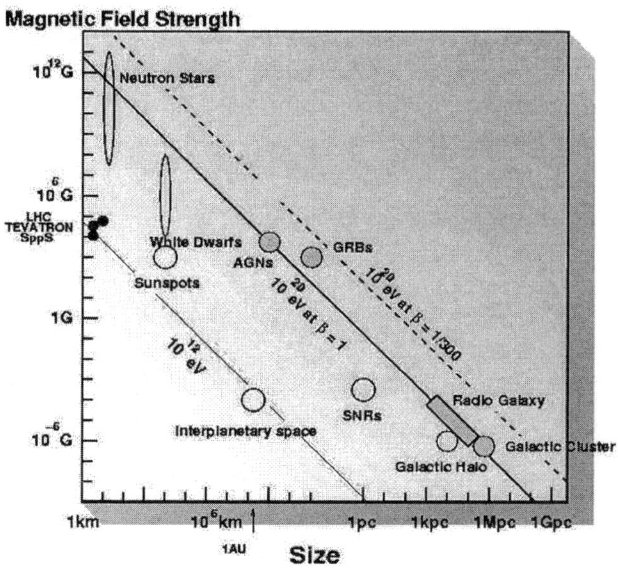

Figure 3. The Hillas plot.

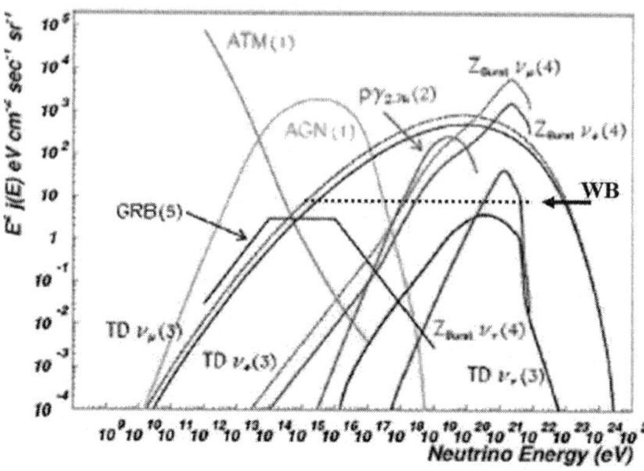

Figure 4. Expected neutrino fluxes.

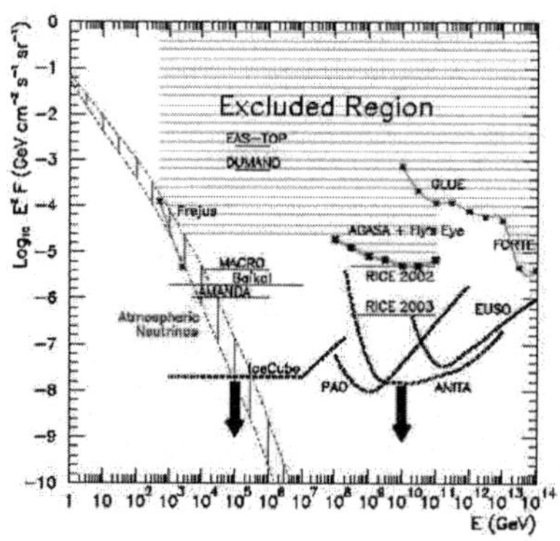

Figure 5. Sensitivities for neutrino detection.

MEASUREMENT OF COSMOLOGICAL PARAMETERS

AMEDEO BALBI

Dipartimento di Fisica
Università di Roma "Tor Vergata"
and INFN, Sezione di Roma II
via della Ricerca Scientifica 1
00133 Roma, Italy
e-mail: balbi@roma2.infn.it

The study of the cosmic microwave background (CMB) has played a crucial role in establishing a new standard of precision in the determination of cosmological parameters. The full-sky, high-resolution maps of the CMB recently produced by the WMAP space mission, have reinforced the evidence in favor of a flat universe, dominated by dark matter and dark energy, and consistent with the predictions of the inflationary scenario. Further high-precision observations of the CMB, such as those that will be carried on by ESA's Planck Surveyor, will further strengthen our knowledge of the Universe.

1. The CMB and cosmological parameters

The cosmic microwave background (CMB) is a powerful tool to investigate the physics of the early Universe and to constrain the parameters of the standard cosmological model. It provides a picture of the Universe when it was only a few hundred thousand years old, at the time when neutral atoms formed and photons decoupled from the matter. The fact that the COBE satellite found the CMB to have a black body spectrum to an astonishing precision (Fixsen et al. 1996) is a clear signature of an early period of matter-radiation equilibrium and a major triumph for the big bang model.

Since the distribution of the CMB photons reflects that of matter at the time of decoupling, any inhomogeneities in the matter density (needed to seed structure formation in the Universe by gravitational instability) must leave an imprint as fluctuations of the CMB temperature. CMB temperature anisotropy was first detected by COBE in the early 90's (Smoot et al. 1992). The fact that the level of anisotropy is very small (about a part in one thousand, corresponding to temperature fluctuations of some tens of μK) simplifies the task of making theoretical prediction of the anisotropy pattern, since linear perturbation theory can be applied.

Most cosmological information encoded in the anisotropy pattern is concentrated at angular scales smaller than about 1 degree on the sky, corresponding to perturbations that were inside the horizon (i.e. in causal contact) before decoupling. On these scales, physical processes in the early Universe were able to leave their imprint on the CMB. The main CMB observable is the angular power spectrum of temperature anisotropy, C_l (see Figure 1). Since each l is related to an angular scale θ on the sky given approximately by $l \sim 180°/\theta$, the power spectrum at high l's probes sub-horizon angular scales at the time of decoupling and carries the imprint of physical processes which occurred in the early Universe. Conversely, low l's basically probe initial conditions in the early Universe (although secondary processes may alter the CMB photon distribution after decoupling).

The detailed shape of the CMB power spectrum is strongly dependent on cosmological parameters. For example, given an initial distribution of density perturbations in the early Universe, the relative height of the peaks in C_l represents a good indicator of the density of baryonic matter in the Universe. On the other hand, the position of the peaks depends on the way the physical scale of sound horizon at decoupling is mapped into an angular dimension on the sky. This mainly depends on the geometry of the Universe: for example, in an open Universe, a certain physical scale at decoupling is seen today under a smaller angle than in a flat Universe. So, the position of the peaks in the CMB angular power spectrum is a good indicator of the total density of the Universe.

In the decade following the release of the COBE results, the experimental efforts focused on measuring the CMB anisotropy at intermediate and small angular scales, that were not accessible to COBE because of its low angular resolution. Several experiments, conducted from 1992 to 1998, either from the ground or from balloon-borne payloads, explored the CMB angular power spectrum in the region between few arcminutes to about one degree. Although each single experiment could only probe a narrow band in l-space, the combined measurements seemed to indicate a rise in the power spectrum at $l \sim 200$.

Thanks to the progress in detector technology, between 1998 and 2000 the experiments TOCO (Miller et al. 1999), BOOMERanG (de Bernardis et al. 2000) and MAXIMA (Hanany et al. 2000) were able independently, for the first time, to clearly resolve the first acoustic peak in the angular power spectrum. BOOMERanG and MAXIMA also produced the first high-resolution (about 10 arcminutes) maps of the CMB, although on small patches of the sky. The detection of the first peak served to support the inflationary scenario, and allowed to measure the total energy density of the universe with unprecedented accuracy. This turned out to be very close to the critical value, $\Omega \simeq 1$, corresponding to a flat universe (Balbi et al. 2000, de Bernardis et al. 2000).

Later, in 2001, the DASI (Halverson et al. 2002), BOOMERanG (de Bernardis et al. 2002) and VSA (Grainge et al. 2003) experiments detected hints of a second acoustic peak in the CMB power spectrum, further strengthening the case for the adiabatic nature of primordial fluctuations. Then, in 2002, the Archeops (Benoit et al. 2003) experiments secured the measurement of the first acoustic peak, and the CBI (Pearson et al. 2003) and ACBAR (Kuo et al. 2002) experiments explored the spectrum at smaller angular scales, measuring the expected damping of primary anisotropy.

Parameter	Symbol	Value
Total density	Ω	1.02 ± 0.02
Baryon density	Ω_b	0.044 ± 0.004
Dark matter density	Ω_m	0.27 ± 0.04
Dark energy density	Ω_Λ	0.73 ± 0.04
Equation of state of dark energy	w	< -0.78 (95% C.L.)
Hubble constant (km s^{-1} Mpc^{-1})	H_0	71^{+4}_{-3}
Age of the Universe (Gy)	t_0	13.7 ± 0.2
Optical depth of the Universe	τ	0.17 ± 0.04
Spectral index of primordial density perturbations	n_s	0.93 ± 0.03

2. The WMAP satellite

The WMAP (Wilkinson Microwave Anisotropy Probe) satellite*, launched by NASA aboard a Delta rocket on June 30, 2001, represents the state of the art of CMB experiments. In many ways, WMAP is a follow-up to COBE. It was designed to make full-sky map of CMB anisotropy by looking at temperature differences in the sky, using differential radiometers in five frequency bands. WMAP scans large regions of the sky in relatively

*http://map.gsfc.nasa.gov

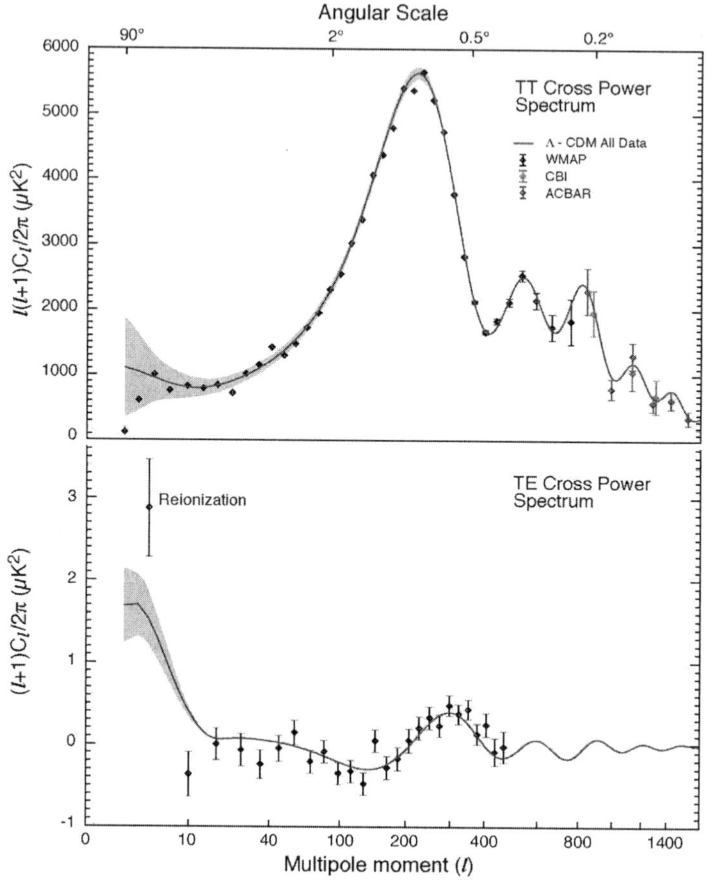

Figure 1. The CMB angular power spectrum. *(Top)*: the temperature power spectrum. The dots are the measurements from the WMAP (Bennett et al. 2003), ACBAR (Kuo et al. 2002) and CBI (Pearson et al. 2003) experiments. The continuous line is the theoretical model which best fits the data. The grey region represents the cosmic variance uncertainty for this theoretical model. *(Bottom)*: the temperature-polarization correlation power spectrum. Note that this is not multiplied by the usual factor of l. (From Bennett et al. 2003)

short times, with a strong cross-linking among observations performed at different times: this is very useful to control systematic effects and correlated instrumental noise. WMAP operates from the L2 Lagrangian point, completing a full sky coverage in a six month period. WMAP detector technology is based on HEMT (High Electron Mobility Transistor) radiometers, passively cooled at about 90 K.

Results of the first year of observations by WMAP (August 2001-August 2002), corresponding to two full-sky surveys, were announced at the beginning of 2003 (see Bennett et al. 2003, and companion papers cited therein). Data collected later are currently being analyzed. The pattern of anisotropy seen in the WMAP maps is consistent with that observed by COBE after four years of observation. WMAP has 30 times better resolution than COBE. When the WMAP map is degraded at COBE resolution, the difference map is below the instrumental noise level.

Figure 1 shows the CMB temperature anisotropy power spectrum measured by WMAP. This is the best currently available measurement of the power spectrum and is cosmic variance limited up to $l \simeq 350$. WMAP results provide an extraordinary confirmation of the theoretical predictions. The presence of at least two acoustic peaks in the power spectrum is evident. The cosmological interpretation of these results lends further support to the standard cosmological model based on big bang plus inflation. A flat universe, with adiabatic, gaussian, scale-invariant primordial density fluctuations is perfectly consistent with the WMAP data. The values of the main cosmological parameters, estimated using the WMAP data, are summarized in Table 1. These values are generally more precise than those obtained with other kinds of observations, and are consistent with them. For example, the baryon density at recombination measured by WMAP is in agreement with big bang nucleosynthesis predictions and measurements of the primordial abundance of light elements (O'Meara et al. 2001, Pettini & Bowen 2001, D'Odorico et al. 2001); the Hubble constant value agrees with the measurement by the Hubble Space Telescope (Freedman et al. 2001); the age of the Universe is consistent with the value from stellar observables (Reid 1997, Hansen et al. 2002); finally, the dark matter content of the Universe is in agreement with the one derived by the large-scale matter distribution (Verde et al. 2002). The low value of the matter density, combined with the fact that $\Omega \simeq 1$, confirms that most of the energy density in the Universe is provided by dark energy, as recently indicated by high-redshift type Ia supernovae observations (Perlmutter et al. 1999, Riess et al. 2001). When combined with other astronomical data (high-redshift type Ia supernovae and the matter distribution inferred from redshift surveys) WMAP observations are able to constrain the equation of state of the dark energy component (see Table 1). The outstanding concordance among completely different kinds of observations testifies the level of maturity reached by cosmology in recent times.

A fraction of the CMB signal is predicted to be linearly polarized. Although the polarized component is expected to be small (about 10% of the total signal) it carries valuable cosmological information. A first detection of CMB polarization, in agreement with theoretical predictions, has been announced by the interferometric experiment DASI (Kovac et al. 2002) and recently confirmed by B2K (Montroy et al. 2005). WMAP detected a correlation between CMB temperature anisotropy and polarization, as expected theoretically (Kogut et al. 2003). This allowed to set limits on the reionization history of the Universe. The integrated optical depth to reionization measured by WMAP is $\tau = 0.17 \pm 0.04$. For a range of plausible models, this corresponds to a reionization at redshift $z_r = 20^{+10}_{-9}$ (95% C.L.) or an epoch $t_r = 180^{+220}_{-80}$ Myr (95% C.L.) after the big bang. Two features of the temperature-polarization correlation measured by WMAP gave further support to the inflationary scenario and to adiabatic primordial density perturbations: the presence of an antipeak at $l \simeq 130$, corresponding to scales that were outside the horizon at decoupling, and the presence of a peak at $l \simeq 300$ that is out of phase with the first peak in the temperature power spectrum (see Fig. 1).

WMAP mission has been approved for 4 years of operation in L2. In the next few years, further data and analysis will provide more and more detailed cosmological information.

3. The Planck Surveyor

Despite its extraordinary achievements, the WMAP mission does not represent the end of the story. Much remains to be told about the CMB temperature anisotropy. On one hand, WMAP angular resolution does not allow to investigate the damping tail of the CMB power spectrum: although the first two acoustic peaks in the spectrum are now accurately resolved, higher l's are affected by large uncertainties. Other experiments, especially interferometers, are starting to unveil the small angular scale details of the anisotropy pattern, but much work needs to be done. On the other hand, WMAP maps are still affected by a non negligible instrumental noise, which strongly reduces the possibility of direct pixel space analyses.

ESA's Planck Surveyor†, planned for launch in 2007, will represent the third-generation CMB space mission. The main product of the Planck mission will be full-sky maps in 9 frequency bands between 30 and 900 GHz. Planck frequency coverage will be the widest ever for a single mi-

† http://astro.estec.esa.nl/Planck

crowave experiment. This is crucial to separate the various components that constitute the observed signal, and will allow the investigation of a large variety of poorly known astrophysical processes, both galactic and extragalactic. Planck will carry on board two different instruments: the HFI (High Frequency Instrument), based on bolometric detectors, and the LFI (Low Frequency Instrument), which uses HEMT radiometers. Exploiting this redundancy and comparison among measurements will be extremely important for the detection and removal of systematics.

Planck's instrumental sensitivity will be several times better than WMAP's. The design of Planck's detectors and optics (a 1.5 m primary mirror and a off-axis secondary, coupled to an array of corrugated horns in the focal plane) will allow to obtain the best possible resolution at each frequency, making it possible to resolve details of a few arcminutes in the sky.

The accuracy of the CMB angular power spectrum measurement by Planck will be limited by cosmic variance and by unavoidable foreground contamination, over the entire range of angular scales relevant to the primary CMB anisotropy, i.e. from $l = 2$ up to $l \sim 1000$, well below the damping scale. This will allow to fully extract the vast amount of cosmological information encoded in the CMB. Planck will be able to measure the cosmological parameters to unprecedented accuracy, minimizing the need of external input from other observations.

The full-sky maps produced by Planck will have a signal-to-noise ratio much larger than 1: this means that Planck's maps will be real pictures of the Universe at recombination. This will allow to accurately investigate physical processes which affect the CMB statistics beyond the angular power spectrum, such as small deviations from gaussianity of the primordial fluctuations, predicted in some theoretical scenarios.

Planck will be the definitive mission for the investigation of CMB temperature anisotropy. On the other hand, Planck detectors were also designed to be sensitive to the polarized component of the CMB signal, although not explicitly optimized for this measurement. This will allow Planck to say something relevant on the new frontier of CMB investigation.

4. Conclusions

Cosmology has developed into a fully mature science. The parameters of the big bang model are now known with great accuracy, and the constraints are expected to get tighter in the future. Inflation has not been falsified, and its main predictions are strikingly consistent with observations. The results obtained using completely different cosmological probes are in remarkable agreement among themselves, as well as with theoretical predictions. Nonetheless, many fundamental questions are still open (see, e.g., Peebles 2003). The pace of experimental and theoretical progress, however, does not seem to be close to a halt.

References

1. Balbi A. et al. 2000, ApJ, 545, L1
2. Bennett C. L. et al., 1996, ApJ, 464, L1
3. Bennett C. L. et al., 2003, ApJS, 148, 1
4. Benoît A. et al., 2003, A&A, 399, L19
5. de Bernardis P. et al., 2000, Nature, 404, 955
6. de Bernardis P. et al., 2002, ApJ, 564, 559
7. D'Odorico S. et al., 2001, A&A, 368, L21
8. Fixsen D. J., Cheng E. S., Gales J. M., Mather J. C., Shafer R. A., Wright E. L., 1996, ApJ, 473, 576
9. Freedman W. L. et al., 2001, ApJ, 553, 47
10. Grainge K. et al., 2003, MNRAS, 341, L23
11. Halverson N. W. et al., 2002, ApJ, 568, 38
12. Hanany S. et al., 2000, ApJ, 545, L5
13. Hansen B. M. S. et al., 2002, ApJ, 574, L155
14. Kogut A. et al., 2003, ApJS, 148, 161
15. Kovac J. M., Leitch E. M., Pryke C., Carlstrom J. E., Halverson N. W., Holzapfel W. L., 2002, Nature, 420, 772
16. Kuo C.-L. et al., 2002, ApJ, astro-ph/0212289
17. Mather J. C., Fixsen D. J., Shafer R. A., Mosier C., Wilkinson D. T., 1999, ApJ, 512, 511
18. Miller A. D. et al., 1999, ApJ, 524, L1
19. Montroy, T. et al. 2003, submitted to ApJ
20. O'Meara J. M. et al., 2001, ApJ, 552, 718
21. Pearson T. J. et al., 2003, ApJ, 591, 556
22. Peebles P. J. E., 2003, astro-ph/0311435
23. Perlmutter S. et al., 1999, ApJ, 517, 565
24. Pettini M. & Bowen D. V., 2001, ApJ, 560, 41
25. Reid I. N., 1997, AJ, 114, 161
26. Riess A. G. et al., 2001, ApJ, 560, 49
27. Smoot G. F. et al., 1992, ApJ, 396, L1
28. Verde L. et al., 2002, MNRAS, 335, 432

THE PRESENT AND THE FUTURE OF COSMOLOGY WITH GAMMA RAY BURSTS

G. GHIRLANDA & G. GHISELLINI
Osservatorio Astronomico di Brera
Via E. Bianchi 46
I–23807 Merate (LC)
E-mail: ghirla@merate.mi.astro.it

Gamma Ray Bursts are among the most powerful astrophysical sources and they release up to 10^{54} erg, if isotropic, in less than few hundred seconds. Their detection in the hard X/γ ray band (at energies $\geq$10 keV) and out to very high redshift ($z\simeq$6.3) makes them a powerful new cosmological tool (a) to study the reionization epoch, (b) to unveil the properties of the IGM, (c) to study the present universe geometry and (d) to investigate the nature and cosmic evolution of the dark energy. While GRBs will surely help to understand the first two issues in the future, the present link between GRBs and cosmology has been made *concrete* by the recent discovery of a tight correlation between their rest frame prompt and afterglow emission properties which allowed their use as *standard candles*.

1. Introduction

Since their discovery in the latest sixties, the observational picture of GRBs have been enriched with new pieces of evidence relative to their emission properties, distance scale and intrinsic properties. Also the theoretical models which pretend to interpret these results were refined (e.g. see [19] for a recent review).

The nature of GRB progenitors, the huge amount of energy emitted during their prompt and afterglow phase, their relativistic nature and their cosmological distance scale makes them one of the most *interdisciplinary* field of research in modern astrophysics. Their study, in fact, requires concepts and methods drawn from the field of stellar evolution, of the physics of compact objects and of relativistic plasmas, of radiative process and they also represent a laboratory to test general relativity and a new tool for cosmology.

The last point gathered the interest of the scientific community since the settlement of the cosmological distance scale of the class of long duration (i.e. $\geq$2 sec) GRBs in 1997 [3]. In fact, the large luminosity of GRBs and their redshift extension ($z \sim 6.3$ up to Sept. 2005 [11]) makes them the brightest distant objects we know of, rising our hopes to test test different cosmological models and to study the nature of dark energy ([8]), i.e. some of the hottest topics of modern cosmology.

2. From the isotropic energy to the collimation corrected energy

Apparently GRBs are all but standard candles: for $\sim$ 30 bursts with measured redshift z and fluence F (integrated over the rest frame 1keV–10MeV energy range) the *isotropic equivalent energy* results,

$$E_{\rm iso} = \frac{4\pi d_L(z)^2}{(1+z)} F \tag{1}$$

with an average value $\sim 10^{53}$ erg and a wide spread over $\sim$3 orders of magnitudes.

However, we know that GRBs are collimated sources with a typical opening angle $\theta_{\rm jet}$ of few degrees [4]. The estimate [18] of the jet opening angle is allowed by the measure of the achromatic jet break time $t_{\rm break}$, i.e. the time (typically between 0.5 and 6 days [13]) when the afterglow light curve steepens. The link between $t_{\rm break}$ and $\theta_{\rm jet}$ is shown in Fig.1. Due to the relativistic beaming of the photons emitted by the fireball, the observer perceives the photons within a cone with aperture $\theta_\Gamma \propto 1/\Gamma$ where Γ is the bulk Lorentz factor of the material responsible for the emission. During the afterglow phase the fireball is decelerated by the ISM and its bulk Lorentz factor decreases, i.e. the beaming angle θ_Γ increases, with time. A critical time is reached when the beaming angle equals the jet opening angle, i.e. $\theta_\Gamma \sim 1/\Gamma \sim \theta_{\rm jet}$, when the entire jet surface is visible any further increase of the beaming angle (due to a decrease of the bulk Lorentz factor) does not increase the visible surface any longer. This time corresponds to a change (steepening) of the afterglow light curve as schematically represented in Fig.1.

The measure of $\theta_{\rm jet}$ allows to correct $E_{\rm iso}$ and derive the *collimation corrected energy*,

$$E_\gamma = E_{\rm iso}(1 - \cos\theta_{\rm jet}) \tag{2}$$

which is however spread over 2 orders of magnitudes and does not make GRBs standard candles.

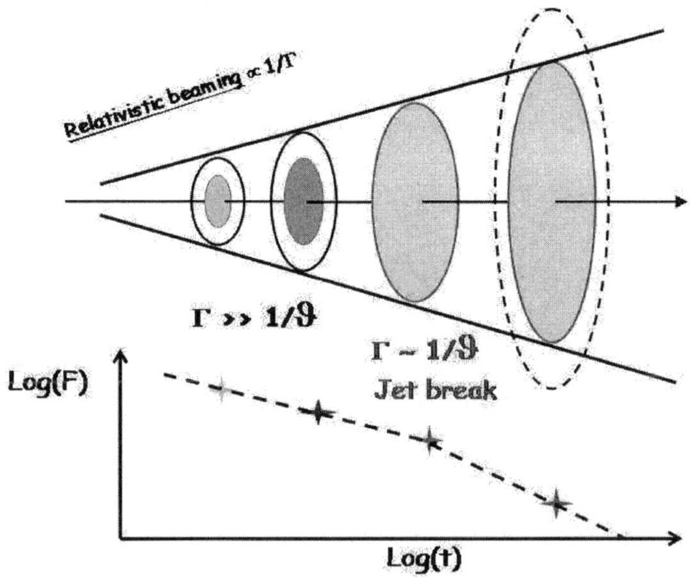

Figure 1. Cartoon of the relation between the jet break time appearing in the afterglow light curve (bottom panel) and the jet opening angle. The shaded surfaces represent the area corresponding to the beaming angle of the photons from which the observer perceives the emitted radiation. When $1/\Gamma \sim \theta_{\rm jet}$ a change in the decay slope of the afterglow light curve appears.

The GRBs whose prompt emission spectrum is known over a wide energy range so that their peak spectral energy $E_{\rm peak}$ (i.e. the peak of the EF_E spectrum) is properly constrained, present a correlation [1] between $E_{\rm peak}$ and $E_{\rm iso}$. With the present sample of 29 GRBs [13] (updated with GRB 051022 - [14]) with accurately measured z, $E_{\rm peak}$ and fluence we find (assuming a standard cosmology with $\Omega_{\rm M} = 0.3$, $\Omega_\Lambda = h = 0.7$):

$$\left(\frac{E'_{\rm p}}{100\,{\rm keV}}\right) = (2.34 \pm 0.01)\left(\frac{E_{\rm iso}}{2.76\times10^{52}\,{\rm erg}}\right)^{0.54\pm0.02} \tag{3}$$

with a reduced $\chi^2 = 4.97$ (27 dof) (Fig.2). The scatter defined by the 27 data points in the $E_{\rm peak}$-$E_{\rm iso}$ plane around the correlation defined in Eq.(3) is well represented by a gaussian distribution with $\sigma = 0.25$ dex.

If we instead correct the isotropic energy for the collimation angle $\theta_{\rm jet}$ as defined in Eq.(2), a *tighter correlation* between $E_{\rm peak}$ and $E_{\rm iso}$ is found [7].

The GRBs for which the jet break time has been measured from the afterglow light curve allows to derive the jet opening angles. To the original

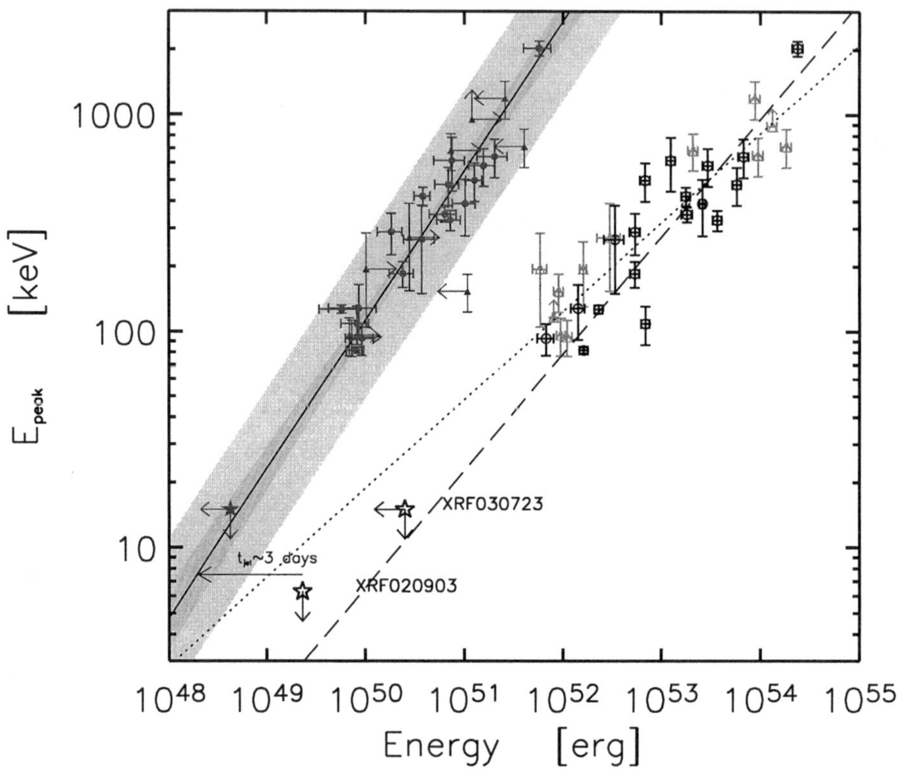

Figure 2. Rest frame peak energy $E_{\rm peak}$ versus isotropic (open symbols) and collimation corrected (filled symbols) energy. The black open circles are the 19 GRBs with measured z, $E_{\rm peak}$ and $t_{\rm jet}$ for which the collimation corrected energy could be computed (red filled circles). Upper/lower limits on either one of the variables are shown by the blue filled triangles. The best fit powerlaw to the red filled circles - Eq.(4) - is represented by the solid line and its uncertainty by the thin yellow shaded region. The large (light orange) shaded region represents the 3σ gaussian scatter of the data points around the correlation. Also shown is the $E_{\rm peak} - E_{\rm iso}$ correlation obtained by fitting the open black data points and the open grey triangles (which are not upper/lower limits) either accounting for the errors on both the coordinates (long dashed line - slope = 0.54 and $\chi^2_{\rm r} = 4.9$ for 27 dof) and by a linear regression (dotted line - slope = 0.4).

sample of 15 bursts of [7] other 4 bursts have been added: GRB 021004, GRB 041006, GRB 050525 and GRB 051022. The correlation defined with this updated sample of 19 GRBs is:

$$\left(\frac{E'_{\rm p}}{100\,{\rm keV}}\right) = (2.82 \pm 0.02) \left(\frac{E_\gamma}{3.72 \times 10^{50}\,{\rm erg}}\right)^{0.69\pm0.04} \quad (4)$$

which gives a better fit (with respect to Eq.(3)), i.e. a reduced $\chi^2 = 1.32$ (17 dof) (Fig.2). Also the scatter defined by the 19 data points in the $E_{\rm peak}$-$E_{\rm iso}$ plane around the correlation defined in Eq.(4) is well represented by a gaussian distribution with $\sigma = 0.1$ dex.

3. Constraints on the cosmological parameters

The tight $E_{\rm peak}$-E_γ correlation makes GRBs standard candles in the sense that, similarly to the stretching–luminosity correlation of SNIa, allows to derive the GRB true energy. In fact, the correlation represents the concrete possibility to use GRBs to test the cosmological parameters and to study the nature of dark energy. In fig.3 we show the constraints on the Ω_M,Ω_Λ parameters obtained with the sample of 19 GRBs.

Although the contours obtained with 19 GRBs are shallow due to the low number of bursts (only 19) compared to SNIa, we can see that GRB constraints are complementary to those defined by SNIa and CMB in the Ω_M-Ω_Λ parameter space. GRBs are detected out to very high redshifts (the present highest detected $z \sim 6.3$ which unfortunately has an unconstrained peak energy) and therefore they can give better constraints in combination with the other cosmological probes.

4. The future of cosmology with GRBs

Two urgent open issues related to the use of GRBs for cosmology are (i) the circularity problem [10] and (ii) the small number of the present sample of GRBs with measured z, $E_{\rm peak}$ and $t_{\rm break}$.

(i) The luminosity distance which allows to compute $E_{\rm iso}$ in Eq.(1) (and E_γ from Eq.(2)) is derived by assuming a cosmological model (defined by the $\Omega = (\Omega_M, \Omega_\Lambda)$ parameters). Therefore, the slope and normalization of the correlation, i.e. $E_{\rm peak} = KE_\gamma^g$, depend on the particular cosmological model adopted, i.e. $K(\Omega)$ and $g(\Omega)$, and we cannot use the correlation found in a particular cosmology (e.g. $\Omega_M = 0.3$,$\Omega_\Lambda = 0.7$) to derive the constraints on the cosmological parameters. A method [7] that circumvents this "circularity" problem is based on fitting the correlation in every cosmology and finding the cosmologies (i.e. the Ω values) which minimize the

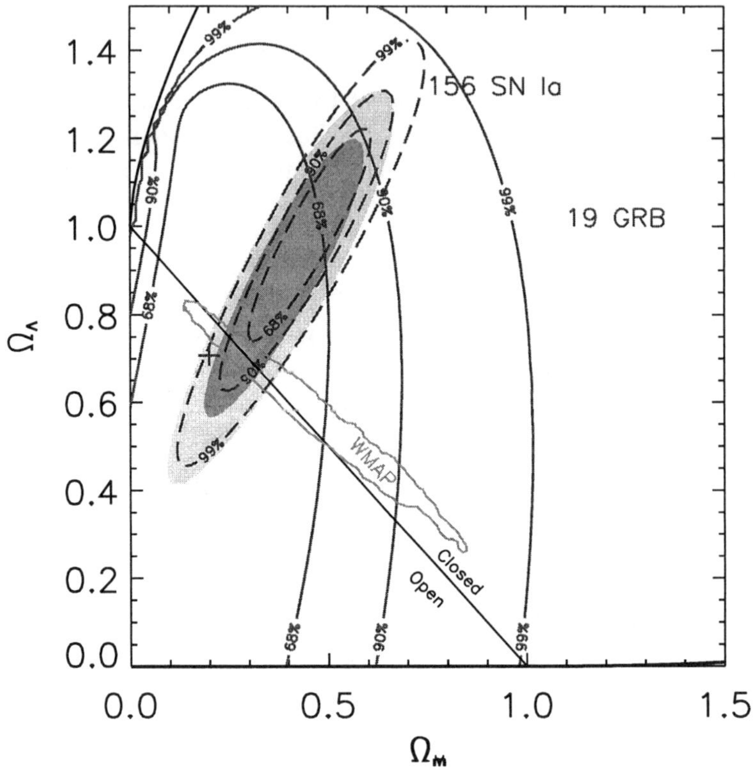

Figure 3. Constraints (1,2 and 3σ) on the cosmological parameters Ω_M,Ω_Λ obtained with the 19 GRBs with z, $E_{\rm peak}$ and $t_{\rm break}$ measured (solid contours). The contours obtained with the 156 SNIa of the Gold sample are also shown (long–dashed contours) and the combined GRB+SNIa contours (shaded regions) are represented. Also shown are the 90% constraints obtained with the WMAP data.

χ^2 of the fit. A more refined method based on Bayes theorem has been proposed and applied [5]. However these methods have the inconvenience that the resulting constraints are less stringent than in the case that the slope g and normalization K of the correlation could be fixed (i.e. cosmology independent). The latter situation can be realized either if the correlation is calibrated with a sample of low redshift GRBs, i.e. at redshifts < 0.1, or if its theoretical interpretation is found. In both cases in fact, the slope and normalization would become independent of Ω. We stress that the requirement of having low redshift GRBs is not so stringent: we estimate [9] that a sample of 12 GRBs with redshift $z \in (0.9, 1.1)$ can be used to calibrate the correlation with a precision better than 1% (i.e. the same accuracy acheived with 12 bursts with $z \in (0.45, 0.75)$). A valid alternative is to find a solid theoretical interpretation of the correlation which would fix the values of K and g independently from the cosmological parameters.

(ii) A simulation of a sample of 150 GRBs shows [9] that the 1σ constraints obtained with GRBs alone might improve of roughly a factor 10 with respect to the present constraints obtained with only 19 bursts (solid contours in Fig.3). However, the increase of the sample of GRBs used in cosmology represents an issue[12] because, for every GRB which is added to the samepl of 19 we need to measure: (i) the redshift z; (ii) the jet break time $t_{\rm break}$ from the X–ray/Optical light curve (this measure requires the monitoring of the GRB light curve up to $\sim$ 10 - depending on the redshift of the source z) (iii) the peak energy of the prompt emission $E_{\rm peak}$

In particular the measure of $E_{\rm peak}$ requires to have a spectrum of the prompt emission of the burst over a wide energy range. In fact, we know that GRB spectra are characterized by a featureless continuum which is often represented [2] by a double smoothly–joined powerlaw model. Typically the EF_E spectrum shows a peak at $\approx$ 300 keV [15] and it has been shown to be extended to very low values (i.e. $\approx$ 20 keV) by the class of X–Ray Flashes [17]. Therefore a detector which can cover the 10 keV–1MeV energy range is required [12] to constrain $E_{\rm peak}$.

Among the presently flying instruments catching GRBs, Swift has the highest detection rate ($\sim$ 1 GRB every 3 days roughly) as it was specifically designed to detect and monitor the burst on–the–fly to pinpoint its position with a high accuracy [6] (to better than $\sim$ 5" in the X–ray band). However, the limited energy range (15–150 keV) of the high energy instrument on board Swift does not allow to constrain the peak energy of most GRBs it detects. However, the IPN satellites and Hete–II [12] have better chances to measure the $E_{\rm peak}$ parameter and, as it appeared from the recent case of GRB 051022, their joint work with Swift might be the key to collect some more GRBs to be added to the presently small sample of GRBs used as standard candles.

Acknowledgments

A. Celotti, C. Firmani, D. Lazzati and F. Tavecchio are thanked for continuous stimulating discussions. The Italian MIUR is thanked for funding

(cofin grant 2003020775_002)

References

1. L. Amati et al. *A&A* **390**, 81 (2002)
2. D. Band et al. ApJ **413**, 281 (1993)
3. S. G. Djorgovski *IAU Circ.*, **6660** (1997)
4. D. A. Frail *ApJ* **534**, 559 (2000)
5. C. Firmani et al. *MNRAS* **360**, L1 (2005)
6. N. Gherels et al. *ApJ* **611**, 1005 (2004)
7. G. Ghirlanda, G. Ghisellini & D. Lazzati, *ApJ* **616** 331 (2004)
8. G. Ghirlanda et al. *ApJ*, **613**, L13 (2004)
9. G. Ghirlanda et al. *MNRAS* submitted
10. G. Ghisellini et al. *Il Nuovo Cimento* in press (2005)
11. N. Kawai et al. *GCN Circ.*. **3937** (2005)
12. D. Lamb et al. *White paper submitted to the Dark Energy Task Force* astro-ph/0507362 (2005)
13. L. Nava et al. *ApJ*, submitted (2005)
14. Olive F. et al. *GCN Circ.* **4131** (2005)
15. R. D. Preece et al. *ApJS* **126**, 19 (2000)
16. A. G. Riess et al. *ApJ* **607**, 665 (2004)
17. T. Sakamoto et al. *ApJ* **629**, 311 (2005)
18. R. Sari *ApJ* **519**, L17 (1999)
19. B. Zhang & P. Meszaros, *IJMPA* **19**, 2385 (2004)

SUPERSYMMETRY BREAKING, EXTRA DIMENSIONS AND NEUTRALINO DARK MATTER

A. M. LIONETTO
Department of Physics and INFN Roma Tor Vergata,
Via della Ricerca Scientifica, 1
00133 Roma, Italy
E-mail: lionetto@roma2.infn.it

We show some phenomenological implications for the dark matter problem of a class of models with deflected anomaly mediated supersymmetry breaking in the context of the MSSM. This scenario can be naturally embedded in a brane world model with one compactified extra dimension. It turns out that in these models the neutralino is still the LSP and so a good candidate as cold dark matter. We found that the neutralino is quite a pure bino in almost all the parameter space. Moreover we computed the thermal relic density and we found wide cosmologically allowed regions for the neutralino.

1. Introduction

Dark matter still remains one of the main unresolved problem in physics. One of the common accepted paradigm is that the solution of the dark matter puzzle involves the existence of an exotic weak interacting massive particle (WIMP). Such a particle has to be found in some extension of the SM of particle physics. It is well known that supersymmetry is an essential ingredient of a consistent theory beyond the SM and the most studied framework is the MSSM, the minimal supersymmetric extension of the SM. In the MSSM the lightest supersymmetric particle (LSP) is usually a neutralino, that is a good candidate as a cold dark matter particle[1]. The pattern of the soft supersymmetry breaking terms* greatly affects the composition and the strength of the dominant interactions of the neutralino. Hence it is very interesting to study the neutralino phenomenology in different supersymmetry breaking scenarios. In more recent developments some of these scenarios can also involve the presence of extra space-time dimensions.

*for a recent review about soft supersymmetry breaking lagrangian see[2]

2. Deflected Anomaly Mediation

The challenge to find a mechanism able to generate a suitable soft supersymmetry breaking lagrangian without any fine-tuning had a considerable attention during at least the last two decades. There are essentially three main classes of solutions to this problem. They differ from each other by the primary "source" that transmits the supersymmetry breaking to the MSSM fields. All these solutions share in common the presence of a hidden sector, that hosts the supersymmetry breaking source, and of a visible sector in which resides the MSSM fields. One of the solutions is the well known gravity mediation (see, e.g.[3]) in which the supersymmetry breaking is vehicled by tree-level Planck suppressed couplings. The scenario with minimal coupled $N = 1$ supergravity is usually called mSUGRA. The mSUGRA models that satisfies the electroweak symmetry breaking constraints constitute the constrained MSSM (CMSSM)[4]. Another solution is gauge mediation (see, e.g. [5]) in which the vehicle are the ordinary gauge interactions. Finally there is the anomaly-mediated supersymmetry breaking (AMSB) scheme (see, e.g. [6] and [8]) in which the supersymmetry breaking is transmitted to the MSSM because of the presence of the superconformal anomaly. Although very natural, the gravity mediated supersymmetry breaking, in order to be phenomenologically viable, implies a fine-tuned forms for the superpotential and Kähler potential[8]. The gauge mediated scheme is not affected by such problem, but on the other hand it very likely predicts the gravitino to be the LSP, which is not the most suitable candidate to constitute cold dark matter.

The anomaly mediated models do not have any of the last two undesirable features, since they usually predict the neutralino to be the LSP and allow the possibility to suppress any direct gravitational coupling between the primary supersymmetry breaking source and the MSSM. This scenario can be naturally embedded in a simple brane world model in a 5-dimensional theory in which the visible sector fields, i.e. the MSSM fields, are confined on a 3-brane. In this kind of models the hidden sector is really a totally sequestered sector, because it resides on a different 3-brane (see fig. 1). One of the main results of Randall and Sundrum[9] is that in order to have a stable brane configuration the space-time geometry between the two branes has to be a slice of AdS_5:

$$ds^2 = e^{-2kr_c\phi}\eta_{\mu\nu}dx^\mu dx^\nu + r_c^2 d\phi^2 \tag{1}$$

where r_c is a "small" compactification radius[9], k is a scale of order the Planck scale and ϕ parametrize the extra (compact) coordinate. In this

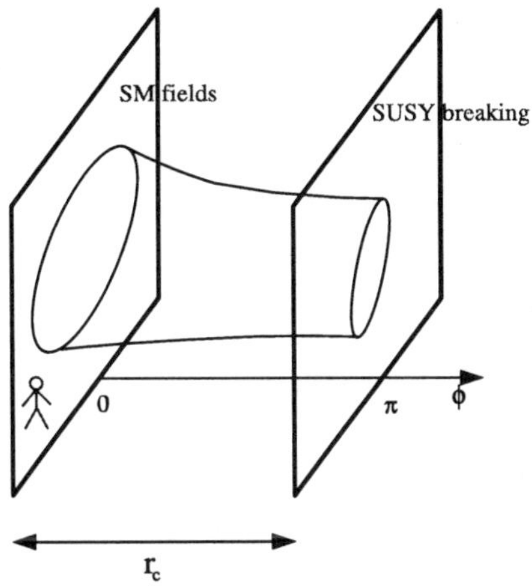

Figure 1. Brane configuration in the Randall-Sundrum model.

way all the massive interaction get suppressed by an exponential factor and only gravity propagates in the bulk. There are no direct contact terms that could lead to phenomenologically dangerous flavor and CP violations effects.

The dynamics in the anomaly mediated sector can be described in terms of a chiral superfield $\varphi = 1 + F_\varphi \theta^2$, the so-called superconformal compensator superfield[7]. This is the only field directly coupled to the visible sector. It is worth to remark that the supersymmetry breaking terms are generated at loop level once the auxiliary field F_φ acquires a VEV[8].

Anyway the "bare" AMSB scheme predicts the uncolored MSSM scalars to be tachyonic and hence it needs for some other mechanism to lift their squared masses to positive values.

A possibility to solve this problem is given by adding another hidden sector[10] in the visible brane containing a gauge singlet chiral superfield $X = (A_X, \Psi_X, F_X)$ directly coupled to N_f copies of messenger chiral superfields Φ_i, with $i = 1, \ldots, N_f$. These fields are assumed to transform under the fundamental and anti-fundamental of the standard MSSM gauge groups as in the usual gauge mediated scenario. Due to the superconformal anomaly, the supersymmetry breaking, originated in the hidden sector, is communicated both to the standard models fields and to the gauge mediated hidden sector fields. We assumed that the supersymmetry breaking effects are transmitted to the gauge mediated sector at tree level and that the superpotential of the gauge singlet $W(X)$ must contain only terms with couplings of positive or vanishing mass dimension:

$$cW(X) = a_1 X^3 + a_2 \langle F_\varphi \rangle X^2 + a_3 \langle F_\varphi \rangle^2 X + a_4 \langle F_\varphi \rangle^3, \tag{2}$$

where a_1, a_2, a_3, a_4 are dimensionless real parameters.

The compensator auxiliary field VEV $\langle F_\varphi \rangle$ (that is coupled to every field in the visible brane) induces nonzero VEVs for both the scalar and auxiliary parts of the X superfield, $\langle A_X \rangle$ and $\langle F_X \rangle$. In this way the messenger superfields Φ_i and $\bar{\Phi}_i$ acquire masses[5] of the order of $\langle A_X \rangle$ and mass splittings of the order of $\sqrt{\langle F_X \rangle}$. Furthermore the presence of the intermediate threshold given by the nonzero VEV of the X superfield will *deflect*† the renormalization group (RG) trajectories of the soft supersymmetry breaking terms off the AMSB trajectory. This deflection is able to eliminate the negative squared masses for the uncolored MSSM scalars, that are present in the minimal realization of the AMSB scenario.

3. Neutralino and Deflected Anomaly Mediation

The phenomenology of the deflected scenario is quite distinctive due to presence of an intermediate energy scale $\langle A_X \rangle = \xi F_\varphi$, where ξ is a new dimensionless parameter, that sets the typical mass of the messenger superfields. It is possible to derive the usual soft supersymmetry breaking terms[10] which depend by the following parameters

$$\langle X \rangle = m(1 + \theta^2 f/m), \quad \langle \varphi \rangle = 1 + \theta^2 F_\varphi \tag{3}$$

where we have defined

$$m = \langle A_X \rangle = \xi F_\varphi \tag{4}$$

$$f = \langle F_X \rangle = d\xi F_\varphi^2 \tag{5}$$

The parameter d indicates how much the RG anomaly mediated trajectory is deflected

$$\frac{f}{m} = dF_\varphi \tag{6}$$

It is possible to show that a superpotential of the type of eq. 2 implies $d > 0$. Thus this scenario is usually termed as positively deflected anomaly mediated[12]. The boundary conditions for the soft terms RG equations are given at the renormalization scale $\mu = m = \xi F_\varphi$. The weak scale M_Z predictions are then obtained by using the usual MSSM renormalization

†for other explicit realization of this class of models see [11] and [12].

group equations at two loop level[13]. To perform the running we used the ISASUGRA RGE package, which is contained in the ISAJET 7.64 package[14]. For the computations of all the relevant quantities at the weak scale we use the DarkSUSY computer code[15]. As we have already seen the soft term expressions are entirely determined by the two mass parameters F_φ and (f/m) and by the dimensionless number N_f. It is then possible to study the phenomenological properties of the deflected model through contour plots in the $(f/m, F_\varphi)$ fixing ξ, which determines the scale at which the boundary conditions are given ($m = \xi F_\varphi$), the ratio between of the two Higgs VEVs $\tan\beta$, the sign of the Higgs μ term and the number of messenger flavors N_f. It turns out that the neutralino is the LSP in almost all the parameter space. This is a consequence of the fact that the fermionic component ψ_X of the hidden sector scalar superfield X has a mass of the order F_φ. Moreover the gravitino mass is $m_{3/2} \sim F_\varphi$[8] and all the soft breaking masses are suppressed by the square of the gauge couplings. Thus neither the fermionic component ψ_X of the gauge singlet superfield nor the gravitino are the lightest supersymmetric particle. The neutralino remains as the main LSP candidate as in mAMSB schemes.

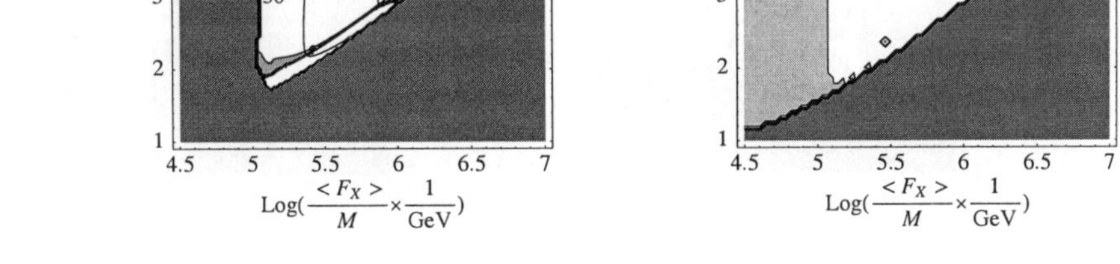

Figure 2. Left panel: neutralino mass m_χ (expressed in GeV) contour plot. The cosmologically allowed regions are in red (WMAP range: $0.09 < \Omega h^2 < 0.13$) and in green ($0.13 < \Omega h^2 < 0.3$) Right panel: exclusion zone (see the text for detailed explanation).

In the left panel of fig. 2 we show the neutralino isomass m_χ (expressed in GeV) contour plot in the plane $(f/m, F_\varphi)$ together with the cosmologically allowed zone: the red region is the one allowed by the WMAP data[16]

$0.09 < \Omega h^2 < 0.13$ while in the green one $0.13 < \Omega h^2 < 0.3$. The dark gray region is the excluded one due to different reasons. In the right panel of fig. 2 we show the details of the excluded region: the blue region is excluded due to the presence of tachyons, the light gray one is excluded by LEP bounds violations while the red one is excluded because there is electroweak symmetry breaking.

In the deflected anomaly scenario the neutralino is a very pure bino in almost all the parameter space. This is the main difference with the mAMSB scenario in which the neutralino is in general wino-like. The range of possible masses for the neutralino is wider than in the well studied mSUGRA case. Unlike than in the latter case it is possible to have a light neutralino of few GeV because the LEP constraints are no more valid here[17] due to the non universality of the gaugino masses at the boundary scale (see eq.7 in 4). Moreover there are regions in which the neutralino has a mass $m_\chi \sim 1$ TeV and also the right cosmological abundance.

In fig. 3 there are other two neutralino isomass contour plots for a different choice of the parameters $\tan\beta$ and N_f. In particular in the left panel $\tan\beta = 50$ and $N_f = 1$ while in the right panel $\tan\beta = 10$ and $N_f = 2$. It

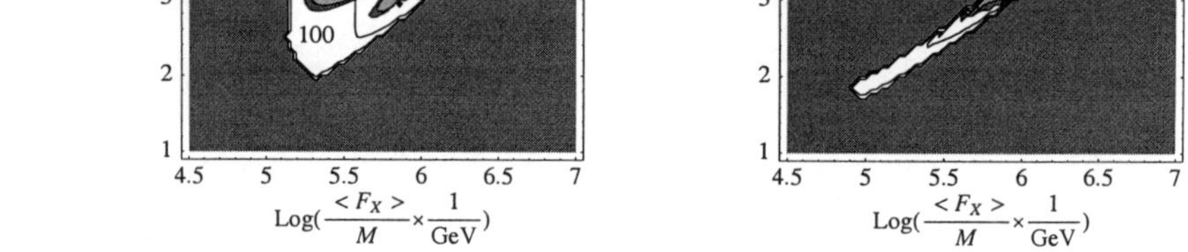

Figure 3. Left panel: neutralino mass m_χ (expressed in GeV) contour plot for $\tan\beta = 50$ and $N_f = 1$ together with the cosmologically allowed regions (see fig.2). Right panel: the same for $\tan\beta = 10$ and $N_f = 2$.

is worth noting that for $N_f > 1$ the parameter space is more constrained with respect to the case $N_f = 1$, i.e. the allowed parameter space is smaller.

4. Conclusions

We presented a peculiar class of models with (positive) deflected anomaly mediated supersymmetry breaking that can be naturally embedded in a brane-world five-dimensional scenario. These kind of models are able to solve the AMSB tachyon problem. Moreover they constitute an interesting alternative scenario in the context of the dark matter problem. We showed that in a deflected anomaly scenario the neutralino is still the LSP and tends in general to be a very pure bino. We found that the neutralino has the right relic abundance to constitute cold dark matter in a wide portion of the parameter space, if compared to other supersymmetry breaking scenarios in which the neutralino is the LSP, as for example the gravity-mediated ones.

Acknowledgments

The author would like to thank Nicolao Fornengo and Stefano Scopel for very useful discussions.

Appendix A. Explicit Expressions for the Gaugino Terms

We show as an example the prediction for the gaugino mass term M_i ($i = 1,2,3$) for a (positively) deflected anomaly mediated scenario. The terms are computed at the renormalization scale $\mu = m$, the typical scale of the messenger fields:

$$\begin{cases} M_1|_{\mu=m} = -\frac{g_1^2(m)}{(4\pi)^2}\left[N_f\frac{f}{m} + (-\frac{33}{5} - N_f)F_\varphi\right] \\ M_2|_{\mu=m} = -\frac{g_2^2(m)}{(4\pi)^2}\left[N_f\frac{f}{m} + (-1 - N_f)F_\varphi\right] \\ M_3|_{\mu=m} = -\frac{g_3^2(m)}{(4\pi)^2}\left[N_f\frac{f}{m} + (3 - N_f)F_\varphi\right] \end{cases} \tag{7}$$

where the g_i ($i = 1,2,3$) are the corresponding gauge couplings. It is interesting to note that these terms are actually the sum of an anomaly mediated and a gauge mediated contribution and that they are not universal. This fact holds only at the messenger scale.

References

1. G. Jungman, M. Kamionkowski and K. Griest, Phys. Rept. **267**, 195 (1996).
2. D. J. H. Chung, L. L. Everett, G. L. Kane, S. F. King, J. Lykken and L. T. Wang,Phys. Rept. **407**, 1 (2005).

3. H. P. Nilles, Phys. Rept. **110**, 1 (1984).
4. G. L. Kane, C. F. Kolda, L. Roszkowski and J. D. Wells, Phys. Rev. D **49** (1994) 6173
5. M. Dine and A. E. Nelson, Phys. Rev. D **48**, 1277 (1993).
   M. Dine, A. E. Nelson and Y. Shirman, Phys. Rev. D **51**, 1362 (1995).
   M. Dine, A. E. Nelson, Y. Nir and Y. Shirman, Phys. Rev. D **53**, 2658 (1996).
   G. F. Giudice and R. Rattazzi, Phys. Rept. **322**, 419 (1999).
6. G. F. Giudice, M. A. Luty, H. Murayama and R. Rattazzi, JHEP **9812**, 027 (1998).
7. E. Cremmer, B. Julia, J. Scherk, S. Ferrara, L. Girardello and P. van Nieuwenhuizen, Nucl. Phys. B **147**, 105 (1979).
   E. Cremmer, S. Ferrara, L. Girardello and A. Van Proeyen, Nucl. Phys. B **212**, 413 (1983).
   A. Pomarol and R. Rattazzi, JHEP **9905**, 013 (1999)
8. L. Randall and R. Sundrum, Nucl. Phys. B **557**, 79 (1999).
9. L. Randall and R. Sundrum, Phys. Rev. Lett. **83**, 3370 (1999).
   L. Randall and R. Sundrum, Phys. Rev. Lett. **83**, 4690 (1999).
10. A. Cesarini, A. Lionetto *in preparation*
11. R. Rattazzi, A. Strumia and J. D. Wells, Nucl. Phys. B **576**, 3 (2000).
12. N. Okada, Phys. Rev. D **65**, 115009 (2002).
13. S. P. Martin and M. T. Vaughn, Phys. Rev. D **50**, 2282 (1994).
14. H. Baer, F. E. Paige, S. D. Protopopescu and X. Tata, arXiv:hep-ph/0001086.
15. P. Gondolo, J. Edsjo, P. Ullio, L. Bergstrom, M. Schelke and E. A. Baltz, JCAP **0407** (2004) 008.
16. D. N. Spergel *et al.* [WMAP Collaboration], Astrophys. J. Suppl. **148**, 175 (2003).
17. A. Bottino, N. Fornengo and S. Scopel, Phys. Rev. D **67**, 063519 (2003).
    A. Bottino, F. Donato, N. Fornengo and S. Scopel, Phys. Rev. D **68**, 043506 (2003).

DARK MATTER AT γ-RAYS

LIDIA PIERI

Department of Physics
Stockholm University,
AlbaNova University Center
SE-106 91 Stockholm, Sweden
E-mail: lidia@physto.se

Indirect detection of Dark Matter through γ-ray astrophysics is a hot topic in the era of new generation Cherenkov telescopes and satellite-borne experiments. We review the particle physics and astrophysics uncertainties which affect the Dark Matter sector and make prediction for the observability of γ-rays produced in Dark Matter annihilation.

1. Introduction

Our ignorance about the nature and distribution of the dark matter (DM) in the universe has been a central topic for discussion among phenomenologists and experimentalists in the last few years.

Analyses of the anisotropies in the cosmic microwave background radiation [1], together with the observed abundance of light elements [2] and the matter power spectrum inferred from galaxy redshift surveys [3] and the Ly-α forest [4] indicate that the Universe contains a significant amount of non-baryonic cold dark matter (CDM).

The hypothesis of a dark matter composed by the lightest supersymmetric particle, which is the neutralino in most cases, and its implications on the prediction of γ-ray flux from DM annihilation has been studied by many authors [5].

In this paper we review the expected γ-ray flux in the frame of an effective minimal supersymmetric model [6], as well as different astrophysical issues that can vary the expected γ-ray flux, following [7,6,8,9].

2. γ-ray flux from dark matter annihilation and experimental sensitivity

Following [6], we write the diffuse photon flux from neutralino annihilation in DM structures as:

$$\frac{d\Phi_\gamma}{dE_\gamma}(E_\gamma, \psi, \theta) = \frac{d\Phi^{\rm SUSY}}{dE_\gamma}(E_\gamma) \times \Phi^{\rm cosmo}(\psi, \theta) \tag{1}$$

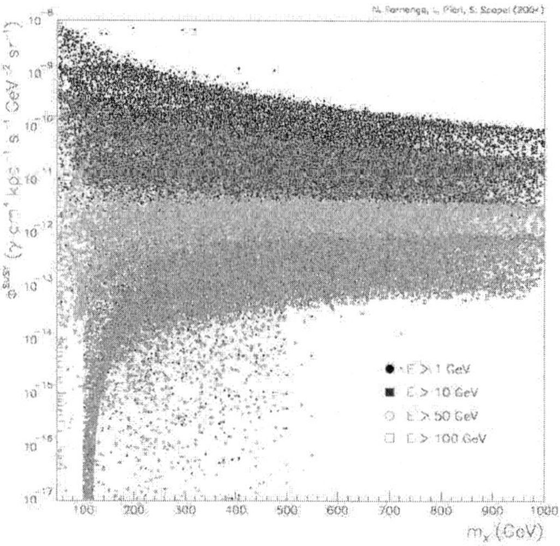

Figure 1. The supersymmetric factor Φ^{SUSY} as a function of the neutralino mass. Different bands show different threshold energies above which the energy spectra have been integrated [6].

The first term, which we will call "supersymmetric factor", deals with the unknown particle physics involved in the DM enigma:

$$\frac{d\Phi^{\rm SUSY}}{dE_\gamma}(E_\gamma) = \frac{1}{4\pi}\frac{\langle\sigma_{\rm ann}v\rangle}{2m_\chi^2} \cdot \sum_f \frac{dN_\gamma^f}{dE_\gamma}B_f \tag{2}$$

where $\langle\sigma_{\rm ann}v\rangle$ is the thermally-averaged neutralino self–annihilation cross–section times the relative velocity of the two annihilating particles, dN_γ^f/dE_γ is the differential photon spectrum for a given f-labeled annihilation final state with branching ratio B_f and m_χ is the neutralino mass. Fig. 1 shows the results on $\Phi^{\rm SUSY}$ obtained in the frame of an effective minimal supersymmetric model, defined as an effective theory at the electroweak scale. The scheme is defined in terms of a minimal number of

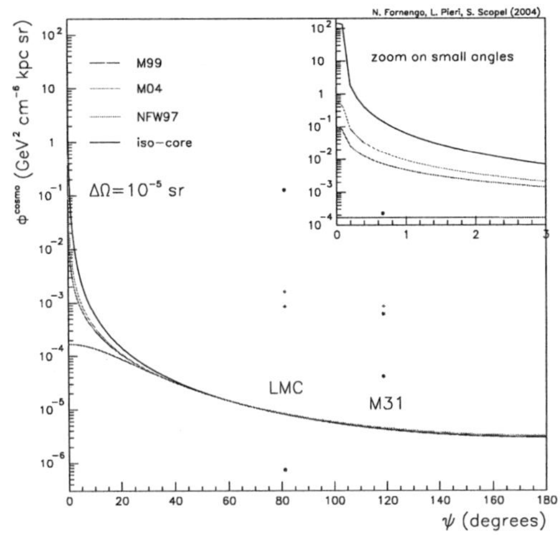

Figure 2. The cosmological factor Φ^{cosmo} for a solid angle $\Delta\Omega = 10^{-5}$ sr for the Milky Way (lines) and for two prominent nearby galaxies (dots), assuming that all the mass is distributed in a smooth DM halo, as a function of the angle of view ϕ from the Galactic Center. The inset shows a zoom at small angles toward the Galactic Center. Different lines show different halo models, both spiky and cored. A constant density central region of radius $r_{cut} = 10^{-8}$ kpc has been used [6].

parameters necessary to shape the essential properties of the theoretical structure of the MSSM and of its particle content, under a number of assumptions which are made at the electroweak scale [6].

The second term, which we call "cosmological factor", is given by the line–of–sight integral and accounts for the astrophysical and cosmological inputs as well as for experimental details:

$$\Phi^{\rm cosmo}(\psi,\theta) = \int_{\Delta\Omega(\psi,\theta)} d\Omega' \int_{\rm l.o.s} \rho_\chi^2(r(\lambda,\psi'))d\lambda(r,\psi') \tag{3}$$

Here, $\rho_\chi(r) \sim r^{-\alpha}$ is the DM density profile, r is the distance from the center of the DM halo, λ is the distance from the observer and $\Delta\Omega(\psi,\theta)$ is the solid angle of observation pointing in the direction of observation ψ and for an angular resolution of the detector θ. The value of the inner slope α is controvert. Neither numerical simulations nor experimental data on galaxy rotation curves [10,11] were able to constrain it. Fig. 2 shows the results on $\Phi^{\rm cosmo}$ for our galaxy and for the LMC and M31, when a smooth DM density profile is assumed, for different choices of α. An inner density core of radius 10^{-8} kpc has been taken into account in order to avoid divergences.

To evaluate the possibility of detecting the γ-ray annihilation line we have studied the sensitivity of the next coming satellite-borne γ-ray experiment GLAST [12] to the differential spectrum of photons expected from DM annihilation, convolved with its energy resolution ΔE=10%.

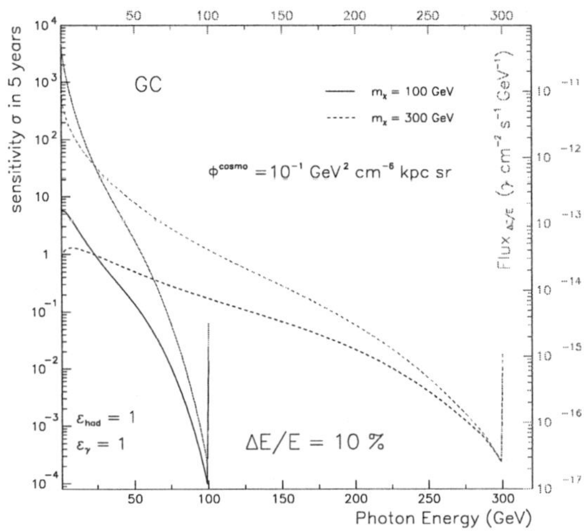

Figure 3. Left y axis (black): Experimental sensitivity of GLAST to the photon flux from DM annihilation coming from the Galactic Center, with $\Phi^{\rm cosmo} = 10^{-1}\,{\rm GeV}^2\,{\rm cm}^{-6}\,{\rm kpc\,sr}$ for $m_\chi = 100\,{\rm GeV}$ (solid line) and 300 GeV (dashed-line). Right y axis (red): Expected flux from the same subhalo as observed by GLAST with an energy resolution of 10%.

We define the sensitivity:

$$\sigma(\Delta E) = \frac{n_{DM}(\Delta E)}{n_{\rm bkg}(\Delta E)} = \frac{\sqrt{T_\delta}\epsilon_{\Delta\Omega}}{\sqrt{\Delta\Omega}} \frac{\int_{\Delta E} A_\gamma^{\rm eff}(E,\theta)[d\phi_\gamma^{\rm DM}/dEd\Omega]dEd\Omega}{\sqrt{\int_{\Delta E}\sum_{\rm bkg} A_{\rm bkg}^{\rm eff}(E,\theta)[d\phi_{bkg}/dEd\Omega]dEd\Omega}} . \quad (4)$$

where T_δ defines the effective observation time, $\epsilon_{\Delta\Omega} = 0.7$ is the fraction of signal events within the optimal solid angle $\Delta\Omega$ corresponding to the angular resolution of $0.1°$, and $A^{\rm eff} = 10^4\,{\rm cm}^2$ is the effective detection area. We optimistically assume that the photon and charged particles' detection efficiencies ϵ_γ and $\epsilon_{\rm ch}$ are 100%, thus we only consider galactic and extragalactic γ-ray background as it is extrapolated by EGRET data at lower energies [13]. Fig. 3 shows the results on the detectability of our Galactic Center (GC), for an inner slope $\alpha = 1$ and 5 years of observations.

Among the astrophysical implications which can change the shape and value of $\Phi^{\rm cosmo}$, we considered the presence of a population of subhalos in the mass range $[10^{-6}, 10^{10} M_{\odot}]$[9], as inferred by recent numerical simulations [14].

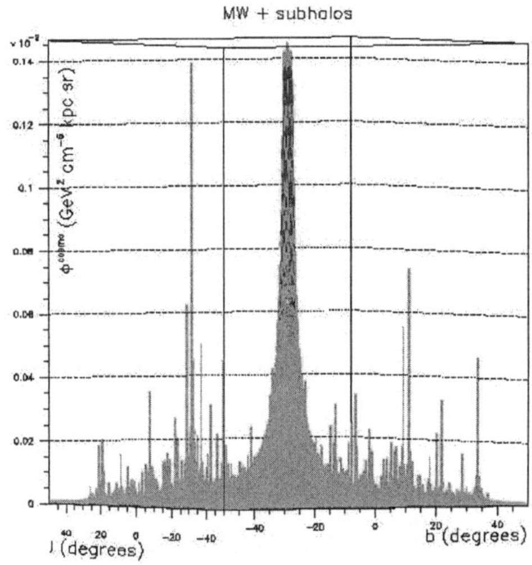

Figure 4. 3D view of the contribution to $\Phi^{\rm cosmo}$ of all subhalos + the Galactic foreground, in a $100^o \times 100^o$ field of view centered on the Galactic Center [9]

All relevant properties of the minihalo population have been modeled in such a way that the number density of subhalos per unit mass at a distance r from the GC is

$$\rho_{sh}(M, r) = AM^{-2} \frac{\theta(r - r_{min}(M))}{(r/r_s^{MW})(1 + r/r_s^{MW})^2} M_{\odot}^{-1}\, \mathrm{kpc}^{-3} \tag{5}$$

We set A such that 10% of the MW mass ($M_{\rm MW} = 10^{12} M_{\odot}$) is distributed in subhalos with masses greater than $10^7 M_{\odot}$ to match the results of [15]. As a result about 53% of the dark mass within our galaxy is not smoothly distributed but contained within $\sim 1.5 \times 10^{16}$ subhalos with masses larger than $10^{-6} M_{\odot}$, corresponding to $\sim 100 \mathrm{pc}^{-3}$ halos in the solar neighborhood. The θ term in Eq. 5 takes into account the effect of gravitational tides which, according to the Roche criterion, disrupt all halos within $r_{min}(M)$ from the GC [16] in an orbital period. The tidal radius $r_{min}(M)$ is an increasing function of the subhalo mass implying that no sub-galactic halos exist within $r_{min}(10^{-6}M_{\odot}) \sim 200\,\text{pc}$ from the GC. $\alpha = 1$ has been used for our Galaxy as well as for each subhalo.

The effect on Φ^{cosmo} is shown in Fig. 4, where the very prominent peak due to the Galactic foreground at the GC position has been artificially truncated to appreciate the subhalos contribution.

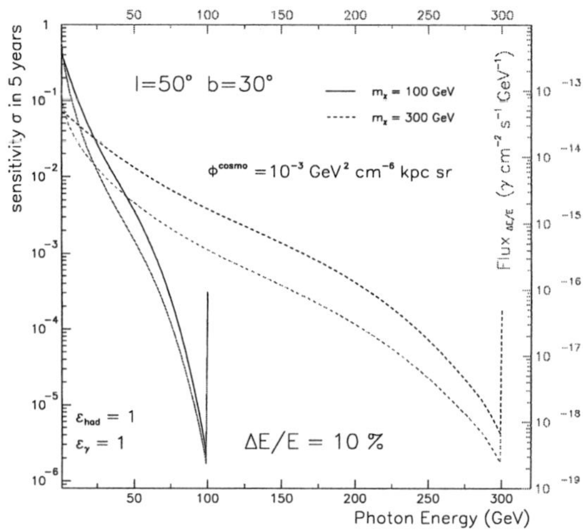

Figure 5. Left y axis (black): Experimental sensitivity of GLAST to the photon flux from DM annihilation coming from a subhalo at $\psi \sim 55^o$, with $\Phi^{\text{cosmo}} = 10^{-3}\,\text{GeV}^2\,\text{cm}^{-6}\,\text{kpc}\,\text{sr}$ for $m_\chi = 100\,\text{GeV}$ (solid line) and 300 GeV (dashed-line). Right y axis (red): Expected flux from the same subhalo as observed by GLAST with an energy resolution of 10% [9].

Fig. 5 shows the sensitivity of GLAST to such a population of subhalos. We simulated a continuous 5 years observation of a fictitious subhalos located at $\psi = 55^o$, assuming that its value of Φ^{cosmo} is the largest value found among all our Monte Carlo realizations.

Another more exotic scenario which could change Φ^{cosmo} involves the postulate of the existence of DM density singularities that arise in irrotational flows in a cold, collisionless medium during the process of structure formation, usually referred to as "caustics" [17,18,19,20]. Following [8], we derive the sensitivity of GLAST to γ-ray emission from neutralino annihilation in the Galactic plane due to the presence of caustic rings, as shown in Fig. 6.

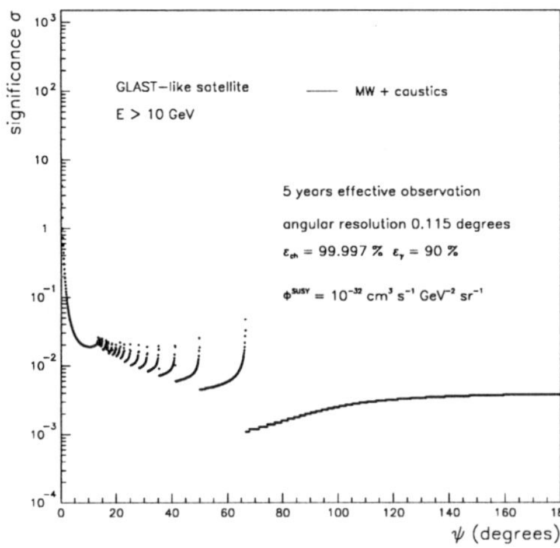

Figure 6. Expected sensitivity of GLAST to γ-ray emission from neutralino annihilation in the Galactic plane due to the presence of caustic rings, at an energy threshold of 10 GeV for a solid angle of 10^{-5} sr.

3. Discussion and conclusion

We have widely explored both the particle physics and the astrophysics involved in the dark matter problem. We predicted for the expected γ-ray fluxes from DM annihilation and extracted the experimental sensitivity of GLAST in the best-case scenario. It turns out that the indirect detection of dark matter through γ-rays afar from the Galactic Center is probably beyond the possibility of planned instruments, unless an unusual steep inner slope for the density profile of substructures is assumed. As far as the Galactic Center is concerned, even in the case of an acceptable $\alpha = 1$ inner slope, it turns to be the best place to look for γ-rays from DM annihilation, although modeling its dramatic environment could prevent us to give a definite answer to the DM enigma, which keeps the game open for both experimentalists and phenomenologists.

Acknowledgments

This proceeding is based on three works done in collaboration with Enzo Branchini, Nicolao Fornengo, Stefan Hofmann and Stefano Scopel.
I acknowledge useful discussions with Lars Bergström, Alessandro Bottino and Fiorenza Donato.

References

1. D. N. Spergel, et al., Astrophys. J. Suppl. **148**, 175 (2003).
2. D. Tytler, et al., Phys. Scr. T **85**, 12, (2000).
3. M. Tegmark, et al., Astrophys. J **606**, 702 (2004).
4. . McDonald, P. et al., astro-ph/0407377.
5. G. Jungman, et al., Phys. Rep. **267**, 195 (1996). L. Bergström, Rept. Prog. Phys. **63**, 793 (2000). G. Bertone, D. Hoooper and J. Silk, Phys. Rept. **405**, 279 (2005).
6. N. Fornengo, L. Pieri and S. Scopel, Phys. Rev. D **70**, 103529 (2004).
7. L. Pieri and E. Branchini, Phys. Rev. D **69**, 043512 (2004).
8. L. Pieri and E. Branchini, JCAP, (2005).
9. L. Pieri, E. Branchini and Stefan Hofmann, astro-ph/0505356, accepted for publication in Phys. Rev. Lett.
10. J. F. Navarro *et al.*, Astrophys. J., **490**, 493 (1997); B. Moore *et al.*, Astrophys. J. Lett. **524**, L19 (1999); P. Ullio *et al.*, Phys. Rev. D **64**, 043504 (2001); J. F. Navarro *et al.*, MNRAS **349**, 1039 (2004); J. Diemand *et al.*, astro-ph/0402267; C. Power *et al.*, MNRAS., **338**, 14 (2003); F. Stoehr *et al.*, MNRAS **345**, 1313 (2003).
11. W. J. G. de Blok *et al.*, MNRAS **340**, 657 (2003); R. A. Swaters, et al., Astrophys.J. **583**, 732 (2003); F. Donato *et al.*, astro-ph/0403206; A. Borriello and P. Salucci, MNRAS **323**, 295 (2001).
12. A. Morselli, et al., in Proc. of the 32nd Rencontres de Moriond, (1997).
13. L. Bergström, et al., Astroparticle Phys. **9**, 137 (1998).
14. J. Diemand, et al., Nature **433**, 389 (2005).
15. S. Ghigna, et al., Astrophys. J, **544**, 616 (2000).
16. E. Hayashi, et al., Astrophys.J. **584**, 541 (2003).
17. P. Sikivie, Phys. Rev. D **60**, 063501 (1999).
18. V.K. Onemli, astro-ph/0401162.
19. P. Sikivie, in *Proceedings of the 2nd International Workshop on the Identification of Dark Matter*, Buxton, England, Sept. 7-11, 1998
20. A. Helmi, S.D.M. White and V. Springel, Phys. Rev. D **66**, 063502 (2002).

POPULATIONS OF SUBHALOS IN COLD DARK MATTER HALOS

E. BISESI

Department of Physics and INFN Udine,
Via delle Scienze, 208, 33100 Udine, Italy
E-mail: bisesi@fisica.uniud.it

We discuss the main properties of substructures in the halos of galaxies. After having introduced a subhalo mass function, we consider a scenario in which the most promising candidates for halo dark matter are weakly interacting massive particles (WIMPs), and give an estimation of the cumulative enhancement in the gamma-ray signal from subhalos of a given mass in the halo of M31. A self-consistent analysis of the dynamical evolution of subhalos is approached, with a particular interest in the description of the effects of mass loss and orbital decay and their implications on the halo merging history.

1. Introduction

The content of matter in the Universe is estimated to be the 27% of its critical density. It is almost universally accepted that most of this matter is non-baryonic. Constraints from primordial nucleosynthesis and cosmic background radiation measurements limit the baryonic content of the Universe to the 4% of the critical density, so the nature of the remaining 23% has yet to be identified. Among the various theories of dark matter, the cold dark matter (CDM) theory provides a remarkably successful description of large-scale structure formation, giving a very good agreement with a large variety of observational data. According with this picture, small objects are the first to form, while bigger structures originate from their progressive clustering. Recently very high resolution numerical simulations confirmed that the dark halos are not smooth structureless objects, but clumpy systems characterized by the presence of a wide population of subhalos.[1],[3],[4] One of the most promising candidates for non-baryonic dark matter is represented by supersymmetric neutralinos. If they exist, they give rise to relic densities in the required range, and are very well motivated in the framework of theoretical extensions of the Standard Model of particle physics. In the context of neutralino dark matter, supercomputer simulations of the formation history of galactic halos have found that the first objects to form should have mass of $10^{-6} M_{\odot}$ and half mass radii of 10^{-2}pc. [1]

2. Mass distribution function

Following [2], we introduce the mass distribution function of subhalos for a given parent halo:

$$N(m|M)dm = A\left(\frac{m}{x\,\beta\,M}\right)^{-\alpha} \exp\left(\frac{m}{x\,\beta\,M}\right)\frac{dm}{x\,\beta\,M}, \qquad (1)$$

which gives the number of subhalos with masses in the range m to $m+dm$, for a parent halo of mass M. This expression gives a very good agreement with recent high resolution dark matter simulations [1],[3],[4],[8] and foresees a total contribution of subhalos of the 18% of the virial mass of the host halo. We note that the x factor in this relation authomatically takes into account the mass loss due to the tidal stripping effect (see section 7.1). Fig. (1) shows the number of subhalos in a halo of mass M as a function of their mass m for a value of the subhalo cutoff of $\beta = 0.39$. We concentrate on the Andromeda galaxy, one of the most promising candidates for dark matter detection in the Local Group, whose main properties are summarized in Table 1. The characteristics of the Draco dwarf spheroidal are also given in this table.

Tab.1 Main properties of M31 and of the Draco dwarf spheroidal [15],[16],[17].

Galaxy	d (Kpc)	r (Kpc)	M(r) (Msun)	rs (Kpc)
Draco	82	7 (rtid)	$6.6 \star 10^8$	0.62 (NFW)
M31	700	295 (rvir)	$1.5 \star 10^{12}$	25 (NFW)

3. Density profiles

In order to give a description of the dynamical evolution of the subhalos, it is necessary to specify the density profiles for both the host halo and the substructures inside the progenitor itself. Among the different models for the initial density distribution, the requirement of consistence with the paradigm of CDM leads to consider the Navarro-Frenk-White profile (NFW):

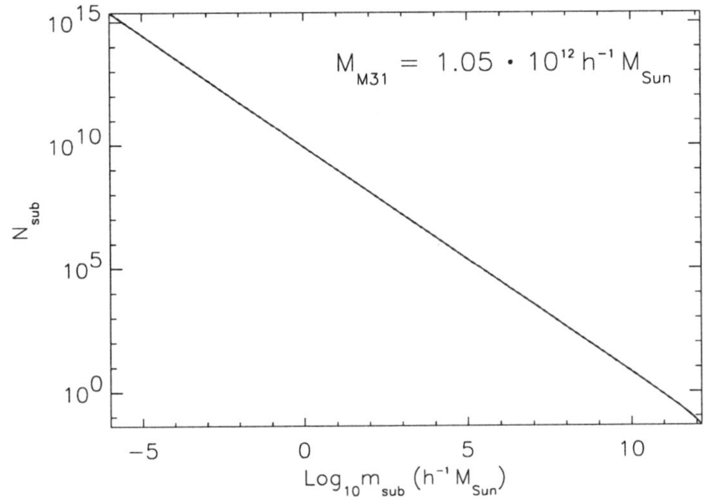

Figure 1. Number of subhalos in the halo of M31.

- **Navarro-Frenk-White** cuspy model [9]:

$$\rho_{NFW}(r) = \frac{\rho_0}{(r/r_s)^\gamma\,(1+(r/r_s))^{3-\gamma}}, \qquad \gamma = 1; \tag{2}$$

On the other hand, the Burkert profile gives the best fits with the rotation curves of some galaxies:

- **Burkert & al.** profile [10]:

$$\rho_{Burkert}(r) = \frac{\rho_0}{(1+(r/r_s))\,(1+(r/r_s)^2)}. \tag{3}$$

While this empirical profile may be justified for the distribution of clumps inside the halo, keeping into accont some global dynamical effects which may have modified the density distribution of the dark matter, it is not realistic to introduce it for the internal distribution of subhalos, which is consistently well described by a NFW profile.
In what follows we will make use of the NFW profile for both the smooth and the clumpy distributions.

4. Gamma-ray flux

Neutralino annihilation in the galactic halo produces both a gamma-ray flux with a continuum energy spectrum and monochromatic gamma-ray lines. Considering a detector with an angular acceptance $\Delta\Omega$ pointing in a direction of galactic longitude and latitude (l, b), the gamma-ray flux from neutralino annihilation at a given energy E is:

$$\Phi_\gamma(E, \Delta\Omega, l, b) = const. \sum_F \frac{<\sigma v_F>}{m_\chi^2} \frac{dN_\gamma^F}{dE} < J(l,b) > (\Delta\Omega) \text{ cm}^{-2} \text{ s}^{-1} \text{ sr}^{-1} \quad (4)$$

for the continuum gamma. The summation term involves the particle physics aspect of the dark matter candidates, while the $< J(l,b) > (\Delta\Omega)$ term is connected with its spatial distribution.
If we assume a spherical dark matter halo in the form of a perfectly smooth distribution of neutralinos, this term is equal to:

$$< J(l,b) > (\Delta\Omega)_{smooth} = const' \frac{1}{\Delta\Omega} \int_{\Delta\Omega} d\Omega' \int_{lineofsight} \rho(L, \psi')^2 dL, \quad (5)$$

where L is the distance from the detector along the line of sight, ψ is the angle between the direction of observation and that of the center of the galaxy and the integration over $d\Omega'$ is performed over the solid angle $\Delta\Omega$ centered on ψ.
In Fig. (2) we plot the density profiles for the dark matter smooth components of Draco and M31 for both the NFW and the Burkert halo models. The values of the PSF are 10^{-5} sr and 10^{-3} sr for the MAGIC and GLAST gamma-ray detectors respectively.

On the other hand, the subhalos contribution to the gamma-ray flux due to neutralino annihilation is given by the following expression:

$$< J(l,b) > (\Delta\Omega)_{clump} = const'' \frac{1}{\Delta\Omega} \int_{\Delta\Omega} d\Omega' \star \quad (6)$$

$$\int_{lineofsight} dL \int_{m_{min}}^{m_{max}} dm \frac{d^2N}{dm\, d^3R} \int_0^{r_{cl}} dr' \;\; 4\pi r'^2 \rho_{cl}^2(m, r').$$

To give an estimation of the enhancement in the neutralino flux due to this component, it is necessary to know also the spatial distribution of subhalos in the galactic halo. The behavior of this function at the present time is not easy to derive, due to the many effects that influence the dynamical evolution of clumps during the halo history. We discuss these effects in detail in Section 7, while in Section 6 we give an estimation of the cumulative enhancement in the gamma-ray signal from subhalos of a given mass.

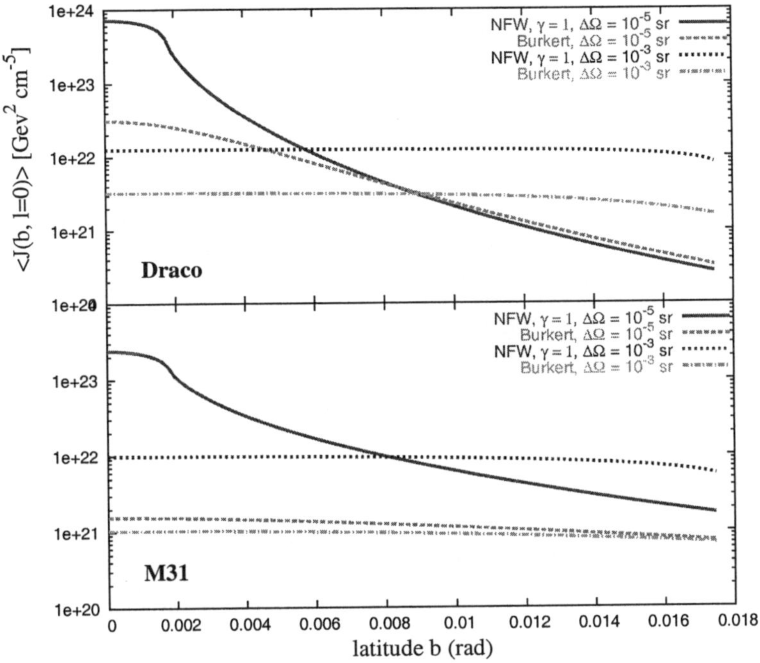

Figure 2. Density profiles for different halo models for Draco and M31.

5. Concentration parameter

Following (Klypin & al., 1999)[7], the concentration parameter is defined as:

$$c_{vir} = \frac{r_{vir}}{r_s}. \tag{7}$$

In Fig. (3) we plot the dependence of c_{vir} on the subhalo mass for M31. We also consider the two toy-models of Bullock & al.[11] and of Eke & al.[12]:

$$c_{vir}^{Bullock}(m, z) = K\,\frac{1+z_c}{1+z}, \tag{8}$$

where z_c is the collapse redshift of the dark matter halo and K is a constant to be fitted to the results of the N-body simulations, and

$$c_{vir}^{ENS}(m, z) = \left(\frac{\Delta_{vir}(z_c)\,\Omega_m(z)}{\Delta_{vir}(z)\,\Omega_m(z_c)}\right)^{1/3}\left(\frac{1+z_c}{1+z}\right). \tag{9}$$

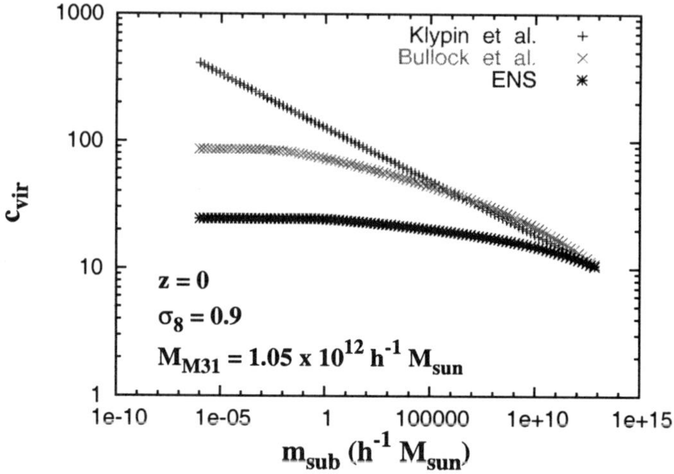

Figure 3. Dependence of c_{vir} on the subhalo mass, following the models of Klypin & al., Bullock & al. and the ENS toy model.

6. Enhancement in the gamma-ray signal.

In order to give a first estimation of the enhancement in the gamma-ray signal due to the presence of subhalos in the halo of M31, we have calculated the ratio between the total flux from the smoothly component and the corresponding contribution for clumps of a fixed mass. Results are plotted in Fig. (4).

7. Dynamical evolution of substructures

Among the main physical processes that drive the evolution of substructures leading to their final distribution, are the mass loss caused by the tidal stripping effect and the orbital decay due to the dynamical frictions. Furthemore, heating due to the close encounters of dark matter halos on their orbits may lead to a considerable mass loss, which would be in any case less efficient than the tidal stripping itself.

7.1. *Tidal stripping*

Considering a subhalo of of mass m and maximum rotational velocity v_{max}, moving at a radius R from the center of a host halo with mass $M(R)$ and maximum rotational velocity V_{max}, its tidal radius r_t is the minimum between the two radii:

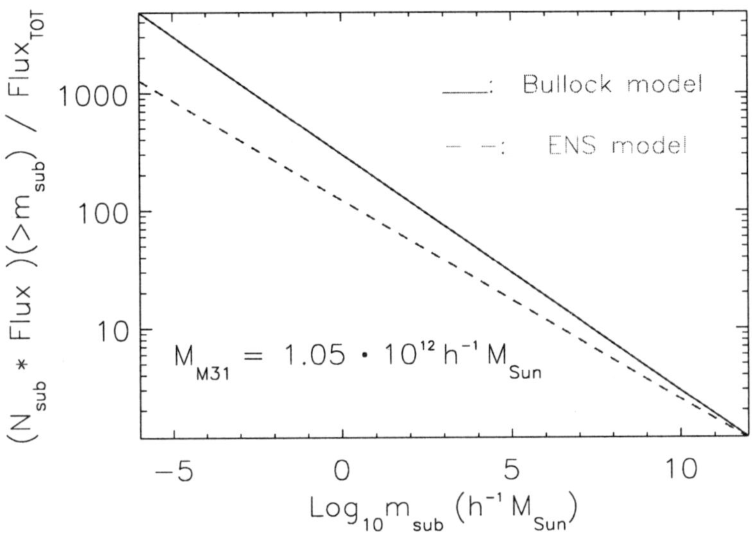

Figure 4. Enhancement in the gamma-ray flux for subhalos of a given mass, for different models of the concentration parameter.

- the radius at which the gravity force of the subhalo equals the tidal force of the host halo

and

- the radius defined by the resonances between the force that the small halo exerts on the dark matter particles and the tidal force by the progenitor.

These two conditions are respectively expressed by:

$$\left(\frac{R}{r_t}\right)^3 \frac{m(r_t)}{M(R)} = 2 - \frac{R}{M}\frac{\partial M}{\partial R} \qquad \text{and} \tag{10}$$

$$\Omega(r)|_{small} = \Omega(R)|_{large}, \tag{11}$$

where $\Omega(r)$ and $\Omega(R)$ are the orbital frequencies of the host halo and of the satellite. By solving these equations numerically, we have obtained the curves drawn in Fig. (5), which show the tidal radius and the mass inside the tidal radius for the whole population of subhalos inside the host halo of M31. These result are derived assuming the NFW profile for both the progenitor and the substructures, and for a concentration parameter as defined in (Klypin & al., 1999)[7].

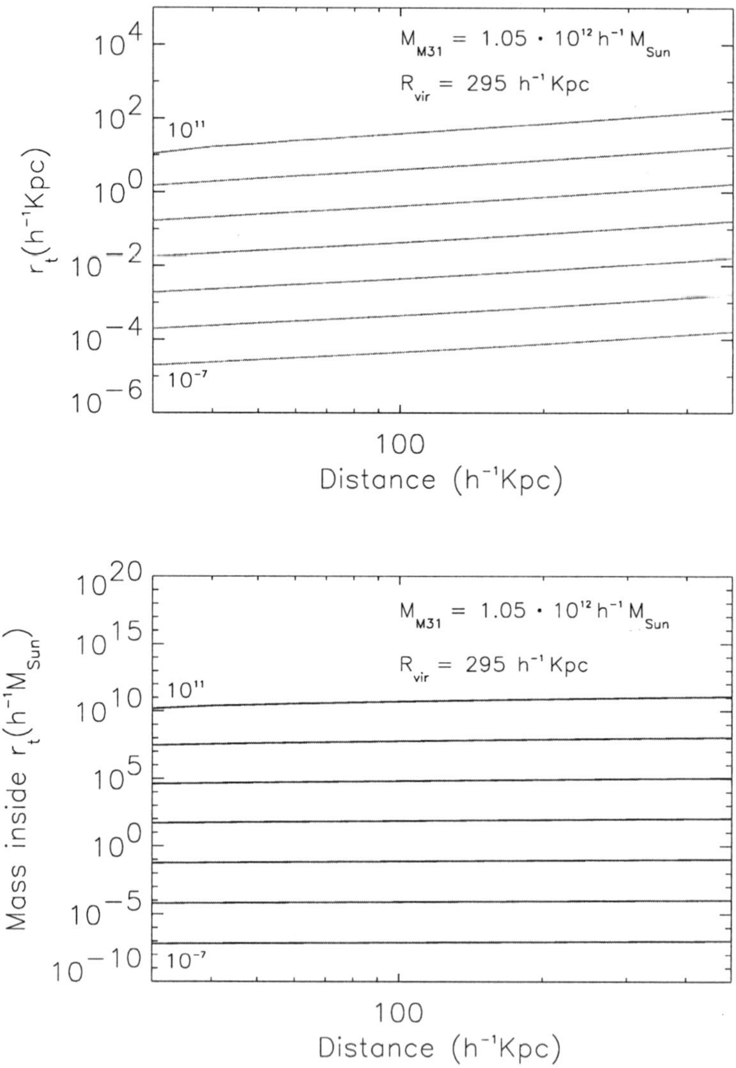

Figure 5. Tidal radius and mass inside the tidal radius of subhalos as a function of their distance from the center of the host halo of M31.

7.2. *Dynamical friction*

The dynamical friction effect causes the orbital decay of the subhalos, making tidal forces longer efficient on them. For a substructure moving on a circular orbit of radius R around the host halo, the rate of the orbital radius decay due to this effect can be estimated using the Chandrasekhar's formula[5], which gives the solution to the differential equation:

$$\frac{dR}{dt} = -\frac{2R}{t_{fric}}\Big[\frac{\partial \ln M(R)}{\partial \ln R} + 1\Big]. \tag{12}$$

Fig. (6) shows the dynamical friction time for different subhalo masses inside the host halo of M31 as a function of the distance of the subhalo from the center of the main halo.

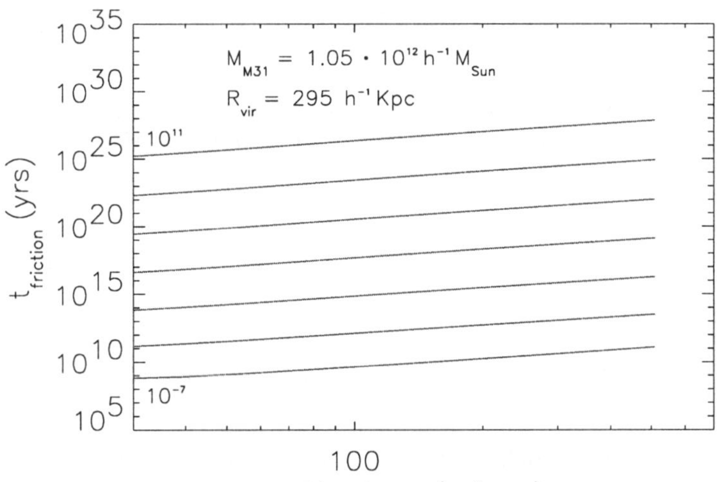

Figure 6. Dynamical friction time for different subhalo masses inside the host halo of M31 as a function of the distance of the subhalo from the center of the main halo.

8. Conclusions

In the present paper we have performed a detailed study of the main properties of substructures of a given mass in the halo of M31, finding a cumulative effect of enhancement in the gamma-ray signal of about a factor 3 for the smaller clumps. This is indicative of the great impact of substructures on dark matter searches and addresses towards the introduction of a more realistic distribution function for subhalos, in which all the spatial effects are taken into account.

References

1. J. Diemand, B. Moore & J. Stadel, *Nature* **433**, 389–391 (2005).
2. A. Vale & J.P. Ostriker, *Mon. Not. R. Astron. Soc.* **353**, 189–200 (2004).
3. J. Weller, J.P. Ostriker, P. Bode & L. Shaw, *Mon. Not. R. Astron. Soc.* **364**, 823–832 (2005).
4. G. De Lucia, G. Kauffmann, V. Springel, S.D.M. White, B. Lanzoni, F. Stoehr, G. Tormen & N. Yoshida, *Mon. Not. R. Astron. Soc.* **348**, 333D (2004).
5. J. Binney & S. Tremaine 1987, *Galactic Dynamics* (Princeton: Princeton Univ., Press), 428.
6. Y. Fujita, C.L. Sarazin, M. Nagashima & T. Yano, *ApJ* **577**, 11–21 (2002).
7. A. Klypin, S. Gottlöber, A.V. Kravtsov, A.M. Khokhlov, *ApJ* **516**, 530 (1999).
8. M. Oguri & J. Lee, *Mon. Not. R. Astron. Soc.* **355**, 120–128 (2004).
9. J.F. Navarro, C.S. Frenk & S.D. White, *ApJ* **490**, 493 (1997).
10. A. Burkert, *ApJ* **447**, L25B (1995).
11. J.S. Bullock & al., *Mon. Not. R. Astron. Soc.* **321**, 559 (2001).
12. V.R. Eke, J.F. Navarro & M. Steinmetz, *ApJ* **554**, 114 (2001).
13. F.C. van den Bosch, G. Tormen & C. Giocoli, *Mon. Not. R. Astron. Soc.* **359**, 1029–1040 (2005).
14. L. Bergström, J. Edsjö, P. Gondolo & P. Ullio, *Phys. Rev. D* **59**, 043506 (1999).
15. N.W. Evans, F. Ferrer & S. Sarkar, *Phys. Rev. D* **69**, 123501 (2004).
16. S. Peirani, R. Mohayahee & J.A. de Freitas Pacheco, *Phys. Rev. D* **70**, 043503 (2004).
17. A. Falvard, E. Giraud, A. Jacholkowska, J. Lavalle, E. Nuss, F. Piron & M. Sapinski, *Astroparticle Physics* **20**, 467–484 (2004).
18. E. Bisesi, M. Mariotti & V. Scalzotto, to be published in the *Proceedings of the 6th International Symposium "Frontiers of Fundamental and Computational Physics" (FFP6), Udine (Italy), Sep. 26-29, 2004*

Part III

Multiwavelength Observations

WEBT MULTIFREQUENCY SUPPORT TO SPACE OBSERVATIONS

C. M. RAITERI AND M. VILLATA*

INAF-OATo, Via Osservatorio 20, 10025 Pino Torinese, Italy
E-mail: raiteri@to.astro.it

The Whole Earth Blazar Telescope (WEBT) is an international collaboration of astronomers interested in blazar variability. Here, we briefly review how it works as well as some among the most interesting results it has reached.

1. Blazars

Blazars are interpreted as AGNs observed at small angles to the jet axis, so that their radiation is dominated by the non-thermal jet emission. The blazar broad-band spectral energy distribution (SED) shows two prominent bumps. The low-energy one is thought to be synchrotron radiation produced by relativistic electrons, while the high-energy one is likely due to inverse-Compton scattering of soft photons by the same relativistic electrons. The peak frequencies of the two components can vary according to the type of blazar considered, but can also change in time for the same object. Indeed, the main feature of blazars is variability. They show flux variability at all wavelengths, from radio to γ-ray frequencies, with time scales ranging from hours (intraday variability - IDV) to years.

2. The Whole Earth Blazar Telescope

The Whole Earth Blazar Telescope[†] (WEBT) is an international collaboration of astronomers interested in blazar variability. About 70 telescopes of various sizes have provided WEBT campaigns with optical and near-infrared data up to now. The distribution in longitude of these telescopes allows the observing task to move from east to west as the Earth rotates.

*For the WEBT collaboration

[†] http://www.to.astro.it/blazars/webt/

Hence, in principle it is possible to obtain 24 hours of continuous monitoring.

Radio telescopes have been providing data at cm wavelengths. The radio telescopes of the University of Michigan (UMRAO, USA) and of the Metsähovi Observatory (Finland) are the historical radio arms of the WEBT. More recently, other radio telescopes have been participating in the WEBT campaigns: the 100 m antenna in Effelsberg (Germany), the 600 m ring RATAN (Russia), and the two 32 m antennas in Medicina and Noto (Italy).

Any scientist interested in blazar variability can submit a proposal to the WEBT President. The President consults with the WEBT officers and if the proposal is judged positively, the start of a new WEBT campaign is announced. The WEBT President also appoints a Campaign Manager, who is usually the campaign proposer. The Campaign Manager is responsible for the success of the campaign. He decides the best observing strategy, and has to take care of the campaign coordination as well as of data collection, analysis, and publication.

A WEBT campaign usually lasts from a few days to several months. From its birth in 1997 up to now, 15 WEBT campaigns have been organized. Sometimes more campaigns were active at the same time. Most of them were supporting satellite observations by CGRO, ASCA, BeppoSAX, RXTE, XMM-Newton, Chandra, and INTEGRAL. Sometimes the high-energy observations were triggered by the discovery of an optical outburst. In a number of cases the WEBT Campaign Manager was the PI of the satellite observation(s). In the other cases, high and low-energy observations were carried out independently.

Data collected during WEBT campaigns are stored in an archive and are available to all the co-authors for one year after the publication of the results; afterwards they become public, and can be obtained upon request to the WEBT President.

3. The WEBT on a bright object: the case of BL Lacertae

BL Lacertae was the target of subsequent WEBT campaigns from 1999 to 2005 (see Refs. 1, and 5–8), in conjunction with pointings by ASCA, BeppoSAX, RXTE and observations by TeV Cherenkov telescopes.

The light curve in the R-band collected by the WEBT collaboration in these years is shown in Fig. 1. The portion of the curve in the period 2002–2005 is still preliminary, since the analysis of the most recent data is still under way[8].

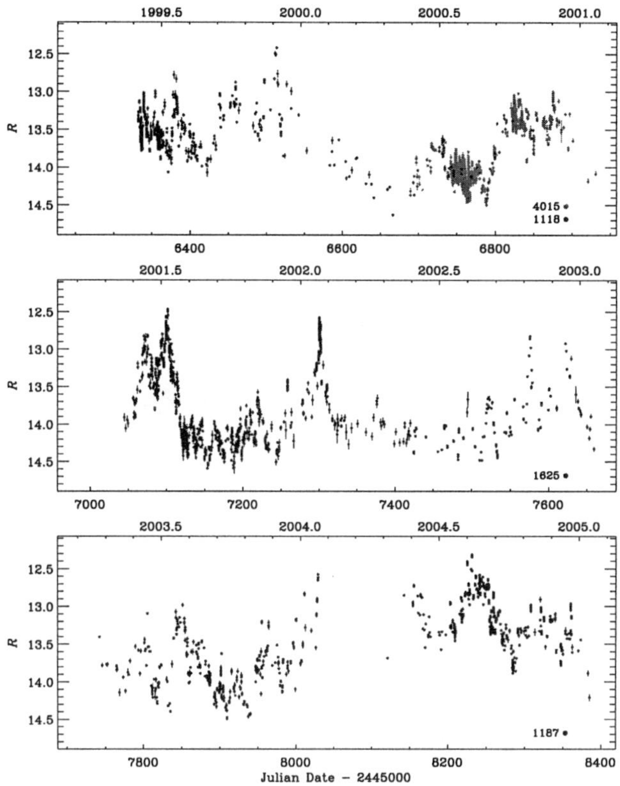

Figure 1. R-band light curve of BL Lacertae in the last years; all data are from the WEBT collaboration; notice the exceptional sampling at the time of the BL Lac 2000 core campaign, in Summer 2000.

The most successful results were obtained during the WEBT campaign BL Lac 2000, in particular during its core period, from July 16 to August 12, 2000 (Fig. 2). In this core campaign data were taken by 19 telescopes located from Japan to Western Canada. The best covered periods show time gaps shorter than 4 hours. The R-band light curve contains 2888 data in 27 days, which means 107 data/day, i.e. a mean time resolution of about 13 min in just one band. If we consider that data were taken also in other bands, the overall sampling is even denser. We notice that time gaps are mainly due to the paucity of observers in the Pacific area.

As one can notice from Fig. 2, the WEBT observations detected many episodes of IDV. The most noticeable ones are enclosed in boxes. As an example, on August 8–9 the source behaviour was followed by 7 telescopes

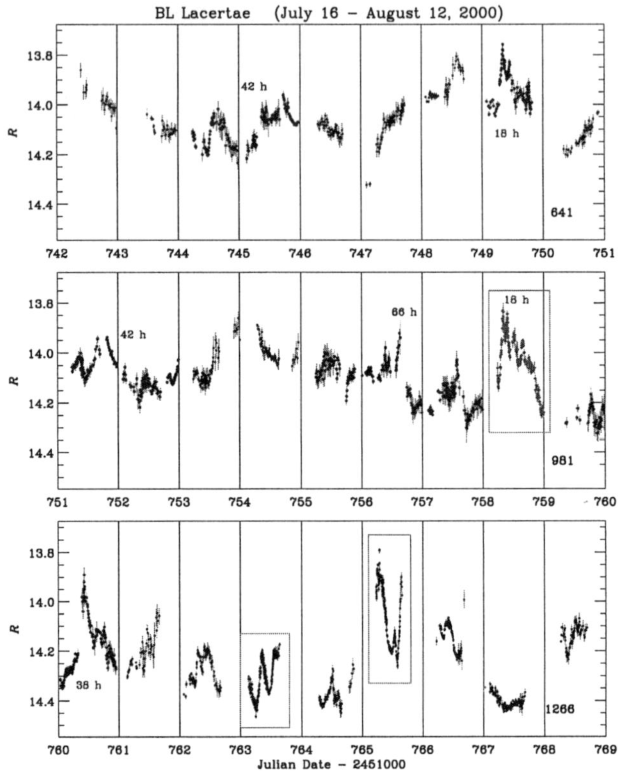

Figure 2. Light curve of BL Lacertae in the R-band during the core period of the WEBT BL Lac 2000 campaign; the source showed noticeable intraday variability.

from Uzbekistan to La Palma. The light curve of this night includes 219 R-band data collected in 10.3 h, which means a mean time resolution of about 2.8 min. A 0.5 mag brightness fall in about 7 h was detected, which was followed by a $\sim$ 0.4 mag brightening in 1.7 h. The WEBT campaign BL Lac 2000 was a great success, leading to unprecedented observing coverage[5]. In total, 15625 optical observations from 22 observatories (24 telescopes) were collected. A huge effort was spent for a careful data analysis and assembling, in order to obtain high-precision light curves.

The analysis of this large amount of data led to the conclusion that an almost achromatic mechanism is causing the long-term modulation of the optical flux base-level, which is likely due to changes of the Doppler factor of geometric origin. On the contrary, fast variations imply spectral changes, and these may be due to intrinsic energetic processes.

Figure 3 shows the BL Lac SED on July 26–27, 2000 compared with the SED obtained in November 1, 2000[1]. In the radio band there were not appreciable spectral variations, which were detected instead in the optical and near-IR bands. But an even more noticeable spectral change was seen in the soft X-ray energy range: in July the BeppoSAX satellite detected a low flux and a hard spectrum; these results were confirmed by simultaneous observations of the PCA instrument onboard RXTE. On November 1st BeppoSAX revealed a brighter state and a steep spectrum, while the spectrum was hard at the higher X-ray energies. PCA data just before and just after this BeppoSAX observation witness a dramatic spectral change in the soft X-rays. Observations at TeV energies by HEGRA could only put upper limits to the source flux.

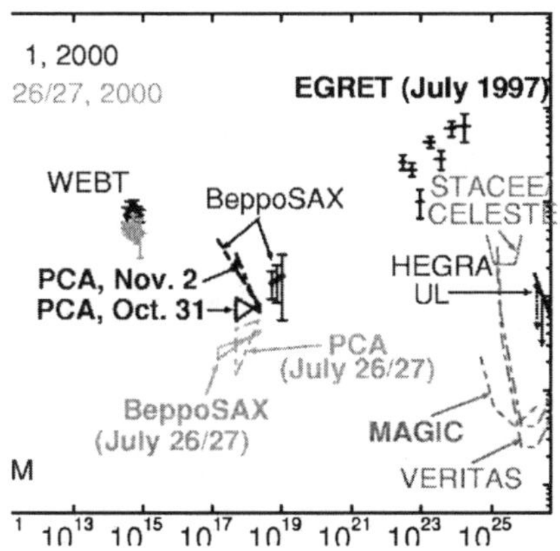

Figure 3. Spectral energy distributions of BL Lacertae in two epochs (July 26–27 and November 1, 2000) during the WEBT BL Lac 2000 campaign; the most dramatic spectral changes were detected in the soft X-rays (from Ref. 1).

Figure 4 shows the optical (top) and radio light curves of BL Lacertae since 1968. The availability of such well-sampled light curves allows a reliable time analysis. In Ref. 7 periodicity analysis was performed by means of three statistical methods (discrete Fourier transform, autocorrelation function, structure function). All of them suggested a characteristic variability time scale of about 8 years for the radio light curves. Moreover, radio outbursts were noticed to have either a hard spectrum or a soft one. The hard radio outbursts seem to be correlated with optical events, radio variations being delayed by about 3 months with respect to the optical ones.

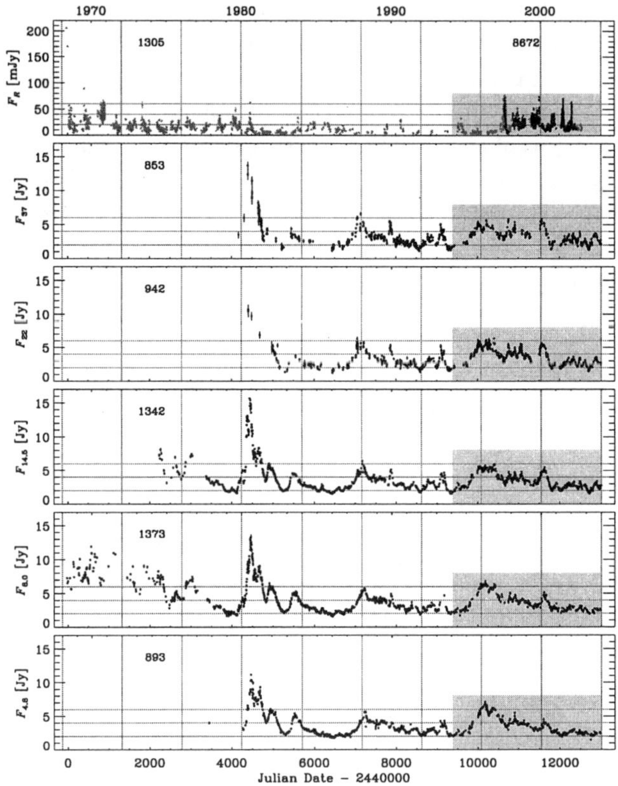

Figure 4. Optical (top) and radio light curves (from 37 to 4.8 GHz) of BL Lacertae from 1968; some radio outbursts are stronger at higher frequencies (hard spectrum), while others imply a higher flux at lower frequencies (soft spectrum).

4. The WEBT on a faint object: the case of AO 0235+16

The intriguing blazar AO 0235+16 was the target of the WEBT first-light campaign in November 1997[2], triggered by the detection of a flaring optical state. A new, long-term WEBT campaign was organized in the two observing seasons 2003-2004 and 2004-2005, to monitor a predicted new radio/optical outburst[3]. The campaign involved radio, near-IR, optical photometric monitoring by the WEBT, VLBA monitoring (15 epochs), optical spectral monitoring at large telescopes (VLT, TNG, NTT), 3 pointings by the XMM-Newton satellite (January and August 2004, January 2005), and dense radio flux monitoring at the 100 m telescope in Effelsberg.

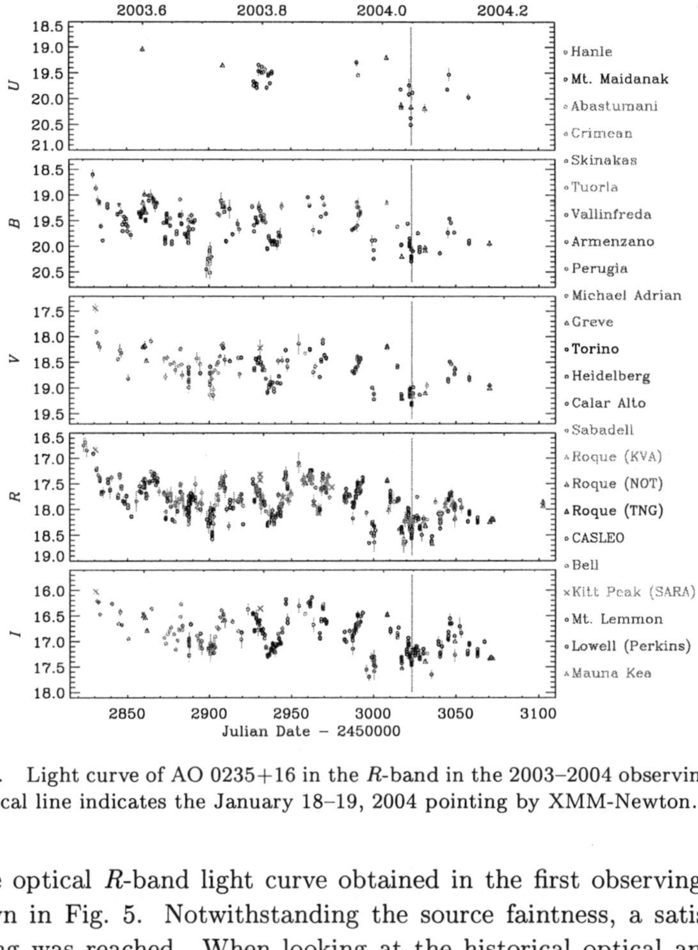

Figure 5. Light curve of AO 0235+16 in the R-band in the 2003–2004 observing season; the vertical line indicates the January 18–19, 2004 pointing by XMM-Newton.

The optical R-band light curve obtained in the first observing season is shown in Fig. 5. Notwithstanding the source faintness, a satisfactory sampling was reached. When looking at the historical optical and radio behaviour of the source, one can recognize little radio activity and just moderate optical variability since 2000[3]. The optical-radio correlation seen during the major outbursts[2] seems to be lost when faint states are considered.

When inspecting the SED built with simultaneous data taken on January 18–19, 2004 during the first XMM-Newton pointing (Fig. 6), a bump appears in the UV–soft-X-ray band, which recalls the "big blue bump" seen

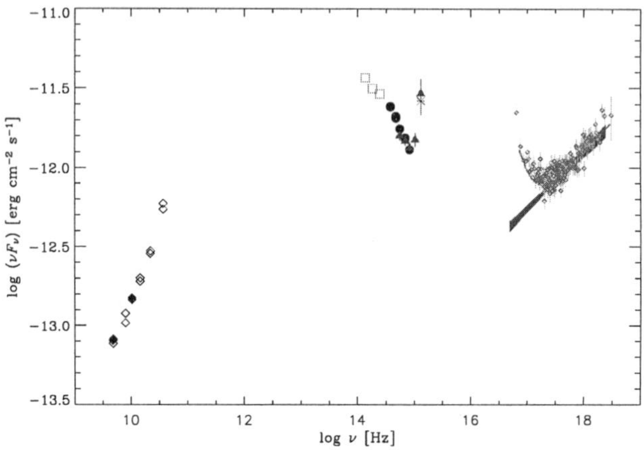

Figure 6. Spectral energy distribution of AO 0235+16 during the January 18–19, 2004 XMM-Newton pointing; the X-ray spectrum is better fitted by a curved model than by a single power law (see Ref. 4); the data suggest a bump in the UV–soft-X-ray domain.

in many AGNs, but may have a different origin[4].

5. Conclusions

The WEBT is the widest collaboration to study blazar variability at low frequencies. WEBTers can provide ground support to high-energy observations with almost continuous sampling and time resolution up to a few minutes for the brightest sources.

References

1. M. Böttcher, A. P. Marscher, M. Ravasio, *et al.*, *ApJ*, **596**, 847 (2003).
2. C. M. Raiteri, M. Villata, H. D. Aller, *et al.*, *A&A* **377**, 396 (2001).
3. C. M. Raiteri, M. Villata, M. A. Ibrahimov, *et al.*, *A&A* **438**, 39 (2005).
4. C. M. Raiteri, M. Villata, M. Kadler, *et al.*, *A&A*, submitted.
5. M. Villata, C. M. Raiteri, O. M. Kurtanidze, *et al.*, *A&A* **390**, 407 (2002).
6. M. Villata, C. M. Raiteri, O. M. Kurtanidze, *et al.*, *A&A* **421**, 103 (2004).
7. M. Villata, C. M. Raiteri, H. D. Aller, *et al.*, *A&A* **424**, 497 (2004).
8. M. Villata, *et al.*, *Astron. & Astrophys.*, in preparation.

REM – THE REMOTE OBSERVATORY FOR GRB ET AL.

EMILIO MOLINARI*
INAF – Osservatorio di Brera, Merate 23807, via Bianchi 46, Italy

The REM (Rapid Eye Mount) observatory was built by a consortium of Italian and international (5%) Institutes in order to have a robotic and fast observing response to GRB alerts. It is located in the Southern Hemisphere (Chile) and can react in 30 sec to a GCN alert. It also allocate time for secondary science proposals and room for future instrumentation.

1. The Telescope

REM is conceived as a link between transient phenomena detected at high energy from space AND Large ground-based facilities on the ground: this is ideal for Gamma Ray Bursts.

This link is needed for transient coordinate determination (in fact often high energy instrument have large error box) and also for a pre-screening of sources for follow-up (which is time consuming and needs large telescopes not always accessible). In any case what we need is: a) coverage in NIR and b) FAST response to alert, and this is how the REM telescope was designed: a fast moving telescope, with alt-az mount, a primary mirror of 60 cm f/8 RC, silver-coated.

It is equipped with 2 Nasmyth foci (one idle) and can move to any point in a cone of 60 deg (the Swift BAT observing cone) in 5 sec, and to any position the sky in 60 sec.

The people who joined this adventure belong to after all a short list of Intitutions: INAF (Osservatorio Astronomico di Brera, Osservatorio Astronomico di Roma, Osservatorio Astronomico di Catania), the University of Milano-Bicocca, The Dunsink Observatory of Dublin and the CEA-SA of Saclay who put together efforts and funds for the telescope and the infrared camera. The later joining of the visible spectrograph and camera gathered also INAF (Osservatorio Astronomico di Bologna, Osserva-

*On behalf of the REM/ROSS team (www.astro-golem.it/projects/rem/)

torio Astronomico di Trieste), the University of Perugia and CNR (IASF Bologna).

The place for the REM location was an issue since the beginning, as it was to follow the Swift alerts as principal source of targets. The uncovered area of the southern hemisphere was the natural choice and the long time connection with ESO turned into Paranal as best possibility. After this ESO itself changed idea and proposed La Silla as better place for REM. After the agreement we started procurement and building for the observatory.

After the initial idea of year 2000, during the following year we placed the order of the telescope with Halfmann GmbH and in 2002 REM was ready and painted. In June the telescope was delivered in Merate for acceptance tests. At the beginning of the next year instrument assembly began in Merate while the dome (Notre Dome) was being built on Cerro La Silla by ESO.

In June 2003 the whole telescope and instrumentation were in the dome and on June 23, 2003, first light with both visible and infrared cameras was obtained.

After first commissioning months the operating problems became clear and we had major failures with Stirling cryocoolers for the infrared camera. After one complete year of increasing difficulties with cooler we decided at the beginning of 2005 to abandon on board cryocoolers and prepared to assemble a new CFC (continuous flux cooling) with liquid nitrogen.

2. The cameras

The optical components of the REM observatory begin before telescope focal plane with a dichroic which is designed to reflect wavelengths shorter than 1 micron (see Fig. 1). The reflected light enters the ROSS visible camera and slitless spectrograph. Ross while the direct light beam will eventually end in the infrared REMIR camera, after passing through an inclined silica plate which can be rotated in order to simulate a dithering of the field, necessary for IR background subtraction. Table 1 summarizes the cameras characteristics.

3. The Dome

The location of the REM dome (Notre Dome) is on the mountain of Cerro La Silla, between ESO 1.5m spectroscopic telescope and ESO 1m telescope. The building is a simple three rooms construction (Fig. 2) with the central one occupied by the telescope with a two parts sliding roof which leaves

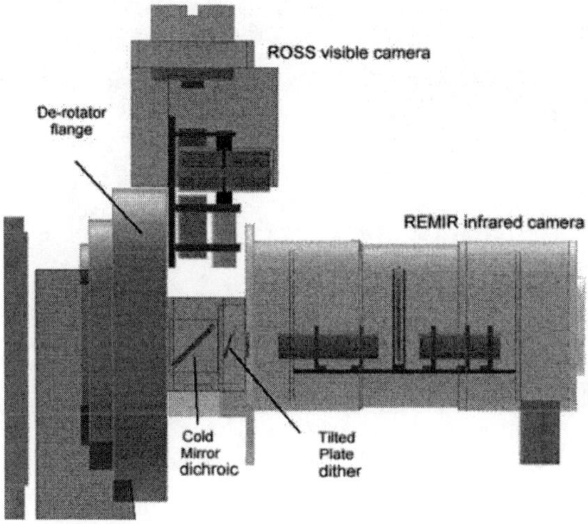

Figure 1. REM Nasmyth instrumentation: after the derotator a cold mirrors splits the incoming beam between visible and infrared wavelengths.

the sky above REM completely clear with no need of rotating domes. The horizon is almost completely free of obstacles, being the 1m telescope the highest point.

Table 1. REMIR and ROSS characteristics.

	REMIR
•	10x10 arcmin FoV
•	1.2 as pixel scale (diff. limited)
•	microns (Z',J,H,Ks)
•	512x512 HgCdTe chip @ 77 Kelvin
•	Wobbling plate for dithering
•	Limiting magnitude (5 sec, S/N=3). J:15.5, H:15.4, K:14.1
	ROSS
•	0.55 arcsec pixel scale
•	30 bins between 0.45-0.9 μm (Amici Prism)
•	10x10 arcmin FoV
•	1024x1024 Marconi CCD in Apogee head
•	Limiting magnitude (5 sec, S/N=3). V:17.1, R:18.5, I:16.5

Figure 2. Notre Dome de la Silla, the REM dome with the sliding roof.

4. REM subsystems

The complexity of the observatory arises from the need of auto-control and auto-organization. The observatory is able to organize a night schedule and perform observation by itself, even without being connected to the external network.

As depicted in Fig. 3, the main brain (REMOS) receives input (i.e. request of targets to be allocated) from two main sources: the GRB alert systems, directly from satellite (IBAS, GCN, REMGRB), and the manual entries from the REM science team for the secondary science targets.

The controller of the observatory looks for suitable conditions of visibility, light and meteo. Also the status of the other domes in La Silla is checked and then the schedule of the observation is started. In any moment, should a new entry be submitted or a GRB alert be received, the schedule is instantly recomputed and made operative.

In case of GRB sequence the two cameras observe in different filters (and also spectroscopically with Amici prism) the area in search of transients (not cataloged) or variable sources. The cross correlation of two cameras should in principle be sufficient to find robust OT detection and alert a follow-up program with larger telescopes.

5. Observation report

Up to now (May 2005) the numbers of triggers followed is still scarce. Due to various circumstances the GRB alerts happened mostly when in Chile

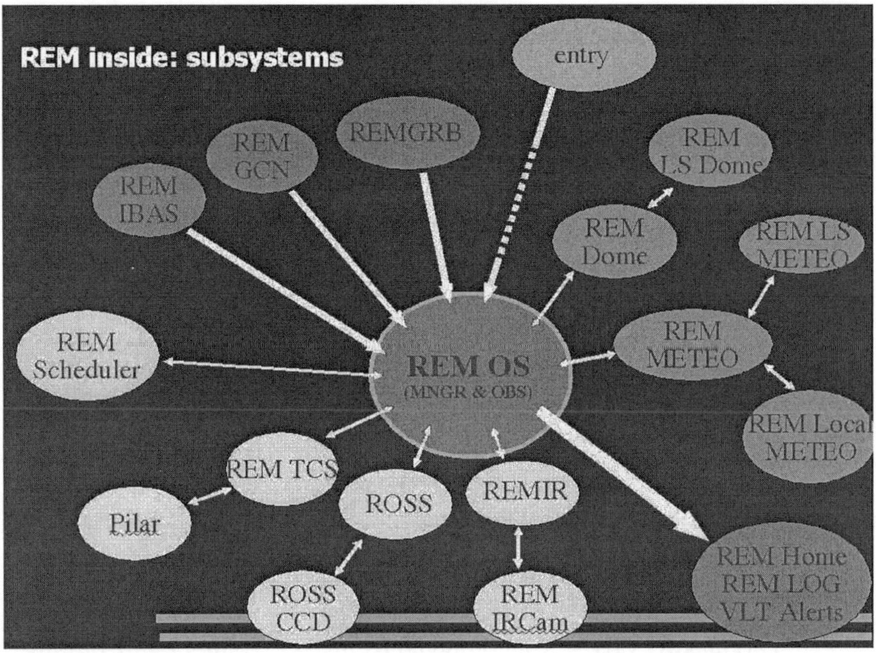

Figure 3. The REM subsystems. Direction of arrows indicates the flow of information within REM observatory.

was daytime or unobservable at too high positive declination. In the cases when it was immediately reachable, 30-35 seconds were measured.

Other topics had more vantage conditions and several observing programs have started with REM.

- GX339-4, a BH candidate, whose bright status was discovered by comparison with the 2mass catalog
- SGR1806-20 and SGR0526-66
- HMXB IGR16318-4848, a High Mass X-Ray Binary. This is the first new gamma source discovered by the INTEGRAL IBIS/ISGRI imager on 2003, January 29. A French-Italian group is looking for variability of the source to check then nature of the collapsed object
- ROSS photometry of open star clusters: col-col diagrams of Trumpler 26, Collinder 347, Trumpler 31.
- Blazars: during monitoring of PKS0537-441 an optical flare was observed on Feb. 2005 followed by a flare in the NIR band.
- 3C 454.3: follow up of the ongoing great outburst. About 50 Opti-

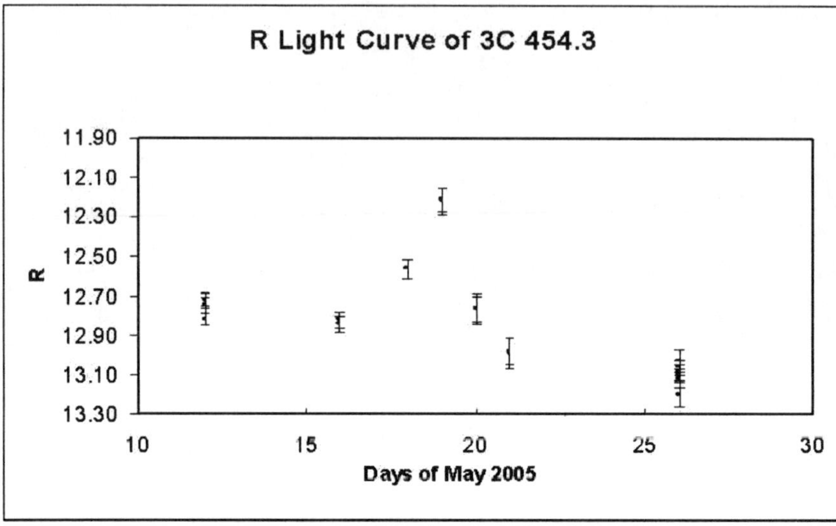

Figure 4. The visible burst of blazar 3C454.3

cal Images (VRI + Amici) and 50 IR images (H filter) were obtained from May 11 to May 17 (Fig. 4)

- Swift Key Project - Observations of a Sample of Blazars. More than 1500 photometric point in Opt/IR already obtained for dozens of targets. Data reduction is ongoing

6. The present and the future

Ongoing operation of the REM observatory can be followed by the REM team and also by interested users via web tools. The present status can be viewed with t-REM-o-meter at http://polaris.merate.mi.astro.it/trem. Current sky and targets over Notre Dome, weather conditions and array/dewar temperature are shown. Also indications on the telescope and camera status are reported. Observational quick look reports cab be looked at via a link in the observation line.

The future of the Observatory is linked to the use of the astronomical community and the future policies for proposal submission and data property. Nowadays the REM Observatory is a PI instrument with access allowed to team members and collaborators, following clear rules for paper authoring.

A new collaboration with Russian Academy of Science will bring a piggy back fast camera for REM. A image intensified CCD camera with 20degress of field of view, 0.13 sec frame rate, reaching visible 11 mag and producing 0.5 Tbyte of data per night. Transient (and among them also satellites and meteorites) are detected, tracked and measured. This will allow to be present also in the very first second of a (bright) optical transient in the aimed portion of sky.

References

1. Covino, S.; Zerbi, F. M.; Chincarini, et al.., "The Multi-frequency Robotic facility REM: first results", *Astronomische Nachrichten*, Volume 325, Issue 6-8, p. 543-548 (2004)
2. Zerbi, F.M., Chincarini, G., Ghisellini, et al., "The REM Telescope: Detecting the Near Infra-Red Counterparts of Gamma-Ray Bursts and the Prompt Behaviour of Their Optical Continuum", *Astronomische Nachrichten.* 322 vol. 5-6, 275-285 (2001)
3. Covino, S., Stefanon, M., Sciuto, G., et al., REM: a fully robotic telescope for GRB observations, Proc. SPIE Vol. 5492, p. 1613-1622, Ground-based Instrumentation for Astronomy; Alan F. Moorwood, Masanori Iye; Eds. (2004)

PLANCK-LFI: OPERATION OF THE SCIENTIFIC GROUND SEGMENT*

F. PASIAN, A. ZACCHEI, M. MARIS, S. FOGLIANI, X. DUPAC,
G. TAFFONI, C. VUERLI, P. MANZATO, F. GASPARO, G. MAGGIO

INAF – Osservatorio Astronomico di Trieste, Via G.B.Tiepolo 11
I-34131 Trieste, Italy

A. GREGORIO

Department of Physics, Università di Trieste, Via Valerio 2
I-34127 Trieste, Italy

D. MAINO, A. MENNELLA, M. BERSANELLI

Department of Physics, Università di Milano, Via Celoria 16
I-20133 Milano, Italy

C. BACCIGALUPI, F. PERROTTA, S. LEACH, F. STIVOLI

SISSA/ISAS, Astrophysics Sector, Via Beirut 2/4
I-34014 Trieste, Italy

R. C. BUTLER, C. BURIGANA, M. MALASPINA, G. MORGANTE,
M. SANDRI, L. STRINGHETTI, L. VALENZIANO, F. VILLA,
N. MANDOLESI

INAF – IASF, sezione di Bologna, Via Gobetti 101
I-40129 Bologna, Italy

G. DE GASPERIS

Università di Roma "Tor Vergata", Via della Ricerca Scientifica 1
I-00133 Rome, Italy

*This paper is written on behalf of the LFI Consortium.

Planck is an ESA mission due to be launched in August 2007 with a payload of two instruments aimed at mapping the microwave sky in 9 frequency bands in the 30-850 GHz range. In this paper, after an introduction describing the basics of the Planck mission and the technical characteristics of the Low Frequency Instrument (LFI), the ground segment activities are discussed, with special emphasis on the monitoring of instrument behaviour, data processing and computational challenges, with a discussion of the foreseen solutions.

1. Introduction

1.1. *The mission*

Planck is a mission of the European Space Agency, ESA, due to be launched in August 2007. The purpose of the mission is mapping the microwave sky with an unprecedented combination of frequency range, angular resolution, sensitivity, sky coverage, calibration accuracy, and freedom from systematic errors. Details are available in [1].

Two Consortia are each in charge of building an instrument and perform the related data processing within a Data Processing Centre (DPC). The Low Frequency Instrument (LFI) is composed of an array of radiometers based on cryogenic HEMT amplifiers cooled to 20 K and working at 3 frequencies (30 - 70 GHz); the PI is N.Mandolesi of INAF/IASF Bologna. The High Frequency Instrument (HFI) is an array of bolometers working at 6 frequencies (100 - 850 GHz), cooled down to 0.1 K; the PI is J.L.Puget of IAS Orsay.

The main scientific goals are to produce all-sky maps of all the major sources of microwave to far-infrared emission, to measure with extreme precision the anisotropies of the Cosmic Microwave Background and to evaluate the main cosmological parameters. In other words, the main scientific steps are the following: (a) imaging the sky at many frequencies; (b) peeling back the layers: detector noise, galactic (dust, synchrotron, free-free), galaxies, clusters, to isolate the CMB signal (while producing a lot of invaluable information for non-CMB science!); (c) recovering the cosmological information.

With respect to previous CMB experiments, Planck has obviously a better sky coverage than balloon-borne or ground-based instruments. With respect to COBE and WMAP, Planck has better resolution and sensitivity, and wider frequency coverage, therefore yielding more accurate maps, and allowing to reduce error bars at higher l's in CMB power spectra (both in temperature and polarisation), revealing all expected acoustic peaks. On the scientific impact of the precision of this measurement, see Balbi et al. [2].

Planck will be launched jointly with the Herschel infrared observatory using the same fairing of an Ariane 5 rocket. The two satellites will separate just after launch and will proceed independently to their final orbits. When in L2, the second Lagrangian point of the Sun-Earth-Moon system, Planck will start spinning in the anti-solar direction at the nominal rate of 1 rpm around L2, with a boresight angle of 85 $\circ$ from the spacecraft axis. The satellite will perform a slew of 2.5 arcmin each hour; this strategy allows the instrument to cover each hour about 60 "rings", which are roughly maximum circles in the sky. The nominal lifetime of the mission is 14 months after reaching L2: the baseline is to perform two full sky surveys. In any case, the consumables on board allow for a possible extension, i.e. at least an additional full sky survey.

1.2. *The instrument*

A detailed description of LFI is available in [3]. The radiometer design is driven by the need to suppress the 1/f-type noise, induced by gain and noise temperature fluctuations in the amplifiers, which would be unacceptably high for a simple total power system. A differential pseudo-correlation scheme is adopted, in which signals from the sky and from a cryogenic blackbody reference load are combined by a hybrid coupler, amplified in two independent amplifier chains, and separated out by a second hybrid. The signals from the two detector diodes ("odd and even samples"), correspond to the sky and the reference load. The radiometer design, however, is such that the 1/f component is highly correlated in the two diodes, and the difference signal is extremely stable and insensitive to 1/f fluctuations. The cryogenic front-end unit is connected to a 300K back-end unit through a set of composite waveguides. By design, all beams have polarisation capabilities. In HFI, about half of the detectors are polarisation-sensitive bolometers, in which a linear grid collects linearly polarised radiation, while the other detectors use spider-web bolometers insensitive to polarisation.

Passive cooling to about 50K is achieved by careful thermal design of the satellite. A 20K stage common to both LFI and HFI is provided by a hydrogen sorption cooler. A closed-loop Joule-Thomson refrigerator provides a 4K stage for further pre-cooling of HFI and cooling of the LFI reference loads; finally, dilution refrigerator provides the operating temperature of the bolometers (0.1 K). The two instruments are integrated in the focal plane of an aplanatic off-axis 1.6-meter aperture double-reflector telescope. A thermal and RF screen, together with the choice of a far-Earth orbit, ensures that residual straylight effects are kept below significance. The combination of LFI and HFI provides full spectral coverage, sensitivity of order $\Delta T/T \sim 10^{-6}$ per resolution element, and a powerful tool to identify and remove systematic effects exploiting simultaneous measurements obtained with two independent technologies.

2. The Scientific Ground Segment and its Operations

2.1. *Monitoring instrument safety and performance*

During the period of satellite visibility (DTCP, 3h per day), the spacecraft and instruments telemetry is dumped to the Ground Station, located in Perth: this includes what has been recorded on board since the preceding DTCP, plus real time telemetry, both housekeeping (HK) and science (SCI). All data are sent to the Mission Operations Centre (MOC), in Darmstadt, where HK telemetry is checked to monitor the safety of the overall system.

Data are then made available to the instrument Consortia through the Data Delivery System (DDS), from where they are retrieved by the first level of the Science Ground Segment (SGS1), run by an Instrument Operation Team (IOT) and hosted at each instrument's Data Processing Centre (DPC). The DPCs have also the possibility of accessing their own computers located at the MOC in order to check in real-time the telemetry being acquired, or recorded telemetry which is still un-consolidated, and therefore still not in the DDS.

The main task of the SGS1 at the LFI DPC is to guarantee safety and health of the instrument under its control, by analysing the HK and SCI telemetry, producing daily and weekly reports in which the instrument performance and trend behavior is defined, and giving feedback to MOC requesting changes in the instrumental setup. Some of the changes (e.g. those dealing with the cooling chain) need to be agreed by both DPCs with the Planck Science Office (PSO), since they are bound to have an impact on the payload as a whole.

2.2. *Data processing*

The processing of the data at the DPC, described in [4] and depicted in Figure 1, aims at producing the final products of the mission (i.e. the frequency maps for all detectors, the maps of astrophysical components, the CMB power spectrum, a catalogue of detected sources). To achieve this result, several processing steps need to be performed. The main processing levels deal with transformation of telemetry into time-ordered streams

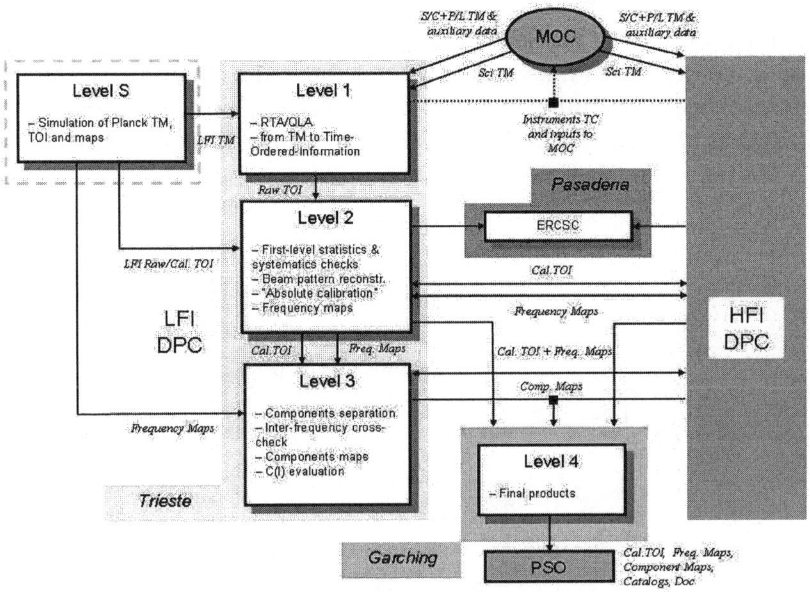

Figure 1. The LFI DPC overall processing scheme.

(Level 1, part of SGS1); then the second level of the Science Ground Segment (SGS2) includes removal of systematic effects, cross-checking between the two instruments and frequency map making (Level 2), separation of the astrophysical components and estimation of the CMB power spectrum (Level 3), archiving and delivery of the final products to ESA (Level 4).

2.2.1. *Systematic effects*

Many issues need to be tackled in data processing, the most critical one being a refined removal of systematic effects, given the high level of precision this experiment requires. This requirement leads to an iterative design of most data processing steps, with the possibility of continuously exchanging data with the HFI DPC.

As an example of removal of a systematic effect, and of feedback of data processing on instrument design, the case of temperature oscillations of the 20K Sorption Cooler can be discussed. Simulations of instrument data, which considered uncontrolled cooler oscillation effects measured during laboratory tests, yielded fluctuations ranging from -2.2 to 2.6 mK; such data can be "de-striped" [11] to reduce the effect, but the impact on the

CMB power spectrum is still high. These simulations led the Sorption Cooler developers at JPL to include a feedback control in the hardware system; the controlled fluctuations were estimated at a level $\sim 10\ \mu$K and, after applying de-striping, the peak-to-peak effect on a pixel in the map is less than 0.2 μK at all frequencies, i.e. quite small compared to the ultimate noise per pixel. The impact in terms of power spectrum is shown in Figure 2.

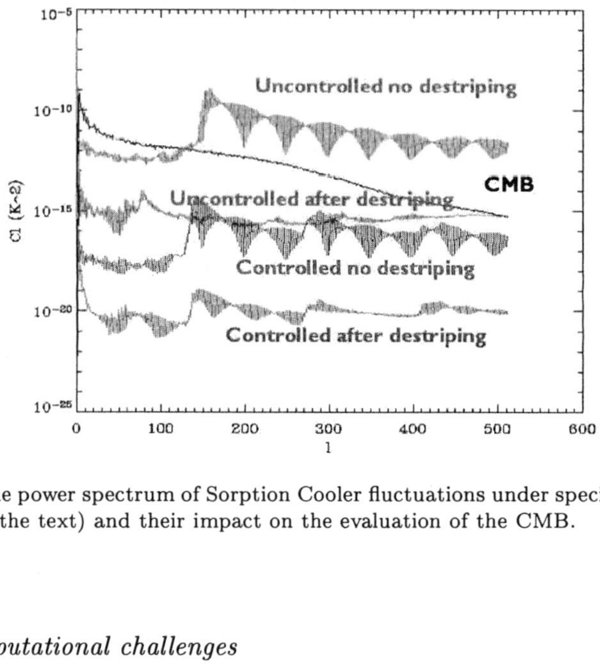

Figure 2. The power spectrum of Sorption Cooler fluctuations under specific conditions (explained in the text) and their impact on the evaluation of the CMB.

2.2.2. *Computational challenges*

Another key point to be mentioned, as pointed out by Bond et al. [5], is the computational complexity of Planck data processing. The general algorithms for mapmaking and component separation perform matrix operations globally on the whole dataset (time series or maps), yielding $O(M^3)$ complexity, where M is the number of elements in the matrix. This was successfully done for COBE (resolution 7°), but for Planck M can be either the pixels of a map ($\sim 10^6$) or the points in the time series ($\sim 10^9$). Furthermore, the knowledge and detail will increase over time, therefore the whole computational chain must be iterated many times. This problem is impossible to solve exactly: simplifying assumptions are needed.

Another demanding data processing step is the estimation of the CMB angular power spectrum: one of the most recent, fast, reliable and unbiased estimator is the so called Montecarlo Apodised Spherical Transform EstimatoR (MASTER) [6] which requires an extensive use of Monte Carlo simulations in order to take into account all issues related to observing strategy (scanning law, sky area covered) and data reduction pipeline. The method has already been applied to Boomerang and WMAP data, and will be particularly demanding in terms of CPU and RAM for Planck.

Given the CMB angular power spectrum and its covariance matrix we can estimate cosmological parameters. The state-of-the-art approach in this respect, implemented by the WMAP analysis team, is based on Monte Carlo Markov Chains to sample in a random fashion (usually with a Metropolis-Hasting algorithm) directly the posterior distribution [7]. The larger the number of chains that is possible to run, the better the convergence of the full set of chains to the "real" posterior distribution. This requires in the order of a hundred of CPUs to solve the problem for Planck with a reasonable efficiency.

2.2.3. *Simulations*

Simulated data are being used to help developing the software of the data processing pipelines, to evaluate before real data are obtained the impact of systematic effects on the scientific results of the mission (as was previously mentioned), and to allow the testing of algorithms needed to solve the critical reduction problems. It is obvious that a large number of runs is needed.

The Planck simulations software [8] is a suite of already-existing applications, developed by institutes involved in the Planck mission which: (a) generate an ideal microwave sky at Planck frequencies by tuning a set of input values (e.g. cosmological parameters, astrophysical foregrounds); (b) "observe" the generated sky by means of instrument models (introducing e.g. noise, systematic effects, non-nominal instrumental behaviour); (c) generate time series (Time Ordered Information); (d) perform simulated on-board processing; (e) generate telemetry. The software is integrated into a coherent pipeline (Level S) by the team at MPA Garching and is resident in a single repository.

A full simulations-processing campaign is currently being carried out for the LFI: the purpose is to evaluate the impact of each systematic effect on the scientific products of the mission, and the ability of the software currently available at the DPC (the so-called Demonstration Model of the

LFI Pipeline) to remove or minimise it. This means that for each systematic effect a full Planck mission is simulated, and the TOI obtained are used to build frequency maps, with and without the processing step aimed at removing the systematic effect itself; the figure of merit to evaluate the effectiveness of the software is the impact of the residual effect on the power spectrum of the CMB.

2.2.4. *Porting to Grid*

Supercomputer facilities have been used in the DPC development. However, considering that the parallelism needed to run Planck simulations and the related data processing is coarse-grained, the solution of using grids to provide adequate computational resources appears to be quite appropriate. The full mission has been simulated on Grid.it, the Italian grid for research, at full resolution for LFI, producing a total amount of TOI of about 1.3 TB for the 22 LFI channels. The mechanism used, foreseen in [9] and described in [10], is shown in Figure 3.

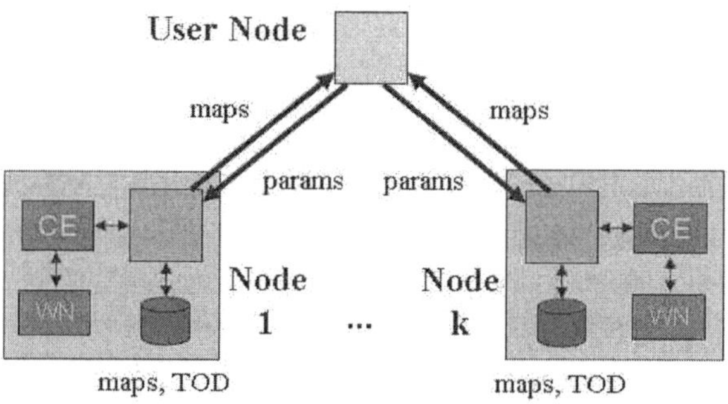

Figure 3. Grid processing for Planck simulations. Cosmological parameters are sent from the User Node to the Grid nodes. In each node an ideal simulated sky is generated, "observed" by the modelled LFI instrument, and processed to remove systematic effects and to produce frequency maps. The maps are then sent back to the user for comparison.

An LFI-led project, approved by the EGEE Generic Applications Panel, has been prepared to massively simulate the Planck mission, also extending the simulations to the HFI channels (74 total, for 5.4 TB of TOI size). Furthermore, a successful application to the DEISA Extreme Computing Initiative was made.

Multiple processing systems are therefore envisaged for the LFI DPC data processing at its main site in Trieste. Standard day-by-day operations (telemetry handling, instrument monitoring, cleaning of data, partial maps) will run on a Beowulf system composed of at least 44 dual-CPU nodes, two for each of the 22 LFI channels in order to allow data re-processing. Creation of the "final" refined data products will require more computing power: in this case, metropolitan or national grid facilities are going to be used, and the CINECA supercomputing centre for tasks requiring large shared memory.

Acknowledgments

The Planck-LFI project is financially supported by National Space Agencies; in particular, the Italian contribution is funded by ASI. Many institutions have contributed to the development of the LFI Data Processing Centre: INAF (OATs, IASF-Bo, IASF-Mi, OAPd), SISSA, University of Milan, University of Tor Vergata Rome, JBO Manchester, IAC Tenerife, University of Cantabria Santander, JPL, US-Planck/LFI, MPA, GADC Geneva, University of Vancouver, University of Helsinki, Metsahovi Observatory, University of Oxford, ESA/RSSD, plus the IDIS Team and many individual contributions.

References

1. J. A. Tauber, Advances in Space Research, 34, 491 (2004).
2. A. Balbi, these proceedings.
3. N. Mandolesi, M. Bersanelli, R. C. Butler, C. Burigana, D. Maino, A. Mennella, G. Morgante, L. Valenziano and F. Villa, in: Experimental Cosmology at Millimetre Wavelengths, AIP Conf. Proc., 616, 193 (2002).
4. F. Pasian and J. F. Sygnet, in: Astronomical Data Analysis II, SPIE proc. 4847, 25 (2002).
5. J. R. Bond, R. G. Crittenden, A. H. Jaffe, L. Knox, Comp. in Sc. & Eng., 1, 21 (1999).
6. E. Hivon, K. M. Górski, C. B. Netterfield, B. P. Crill, S. Prunet, F. Hansen, ApJ, 567, 2 (2002).
7. L. Knox, N. Christensen and C. Skordis, ApJL, 563, L95 (2001).
8. M. Reinecke, K. Dolag, R. Hell, M. Bartelmann, T. Ensslin and M. Ash-down, A&A (2005), in press, astro-ph/0508522.
9. R. Smareglia, F. Pasian, C. Vuerli and A. Zacchei, in Astronomical Data Analysis Software and Systems XIII, ASP Conf. Proc., 314, 674 (2004).
10. G. Taffoni, D. Maino, C. Vuerli, G. Castelli, R. Smareglia, A. Zacchei, T. Ensslin and F. Pasian, IEEE Trans. on Computers (2005), in press.
11. D. Maino, C. Burigana, K. M. Górski, N. Mandolesi, M. Bersanelli, A&A, 387, 356 (2002).

INTEGRAL THREE YEARS LATER

L. FOSCHINI, G. DI COCCO, G. MALAGUTI
INAF/IASF, Sezione di Bologna
Via Gobetti 101, 40129 - Bologna (Italy)
E-mail: foschini, dicocco, malaguti @bo.iasf.cnr.it

The ESA *INTErnational Gamma-Ray Astrophyisics Laboratory* (*INTEGRAL*) is composed of two main instruments (IBIS and SPI) for the γ−ray astrophysics (20 keV – 10 MeV), plus two monitors (JEM-X and OMC) for X-ray and optical counterparts. It was launched on October 17th, 2002, from the Baikonur cosmodrome (Kazakhstan) and it will be active at least up to 2008. A selection of its main scientific contributions obtained to date are presented.

1. Introduction

The *INTE*rnational *G*amma-*R*ay *A*strophysics *L*aboratory (*INTEGRAL*)[1] is an observatory of the *European Space Agency* (ESA). Launched in 2002, with a nominal duration of 2 years, it has already been extended up to 2008. *INTEGRAL* has been built by the ESA member states with the participation of United States, Russia, Czech Republic, and Poland. *Alenia Spazio* (Italy) has been appointed by ESA as prime contractor for the design, integration, and testing of the satellite[2]. The *Mission Operation Centre* (MOC) at Darmstadt (Germany) is continuously in touch with the satellite through two ground stations, one of ESA located at Redu (Belgium) and the other of NASA at Goldstone (USA). The *INTEGRAL Science Operation Centre* (ISOC) was initially at ESA ESTEC (The Netherlands), and in 2005 moved to ESAC in Spain[3]. The *INTEGRAL Science Data Centre* (ISDC) is located at Versoix in Switzerland[4].

Two are the main instruments on board: SPI (*SPectrometer on Integral*)[6] and IBIS (*Imager on Board Integral Satellite*)[5]. The first is dedicated to high-energy resolution spectroscopy (2.5 keV at 1.3 MeV) and it is composed of 19 Germanium detectors operating in the 20 keV – 8 MeV energy range. The second is an imaging telescope optimized for high-resolution (12′) wide field of view (FOV, 29° × 29°) point source reconstruction with moderate energy resolution (8% at 100 keV and 10% at

1 MeV). IBIS is composed of two layers: ISGRI (*INTEGRAL Soft Gamma-Ray Imager*[7]), a 128 × 128 pixels CdTe detector organized in 8 modules operating in the 20 keV – 1 MeV energy range, and PICsIT (*Pixellated Imaging CaeSium Iodide Telescope*[8]), a 64 × 64 pixels CsI detector organized in 8 modules operating in the 175 keV – 10 MeV energy range.

These instruments are complemented by two monitors: JEM-X (*Joint European Monitor X-rays*[9]), operating in the 3 – 35 keV energy band with angular resolution of $3'$, and OMC (*Optical Monitor Camera*[10]), for optical study in the Johnson V band with magnitude limit 18.

IBIS, SPI, and JEM-X detectors are all coupled with tungsten coded-masks: SPI and JEM-X make use of a hexagonal uniformly redundant array (HURA), while the IBIS mask is a modified uniformly redundant array (MURA). Specific mathematical operations (deconvolution) are needed to obtain images, spectra, and lightcurves, and are described in Goldwurm et al.[11] and Gros et al.[12] for IBIS, in Skinner & Connell[13] and Strong[14] for SPI, and in Westergaard et al.[15] for JEM-X.

Since *INTEGRAL* is an observatory, most of its time is dedicated to the astronomical community, through annual announcements of opportunity (AO). However, part of the observing time is reserved to the Instrument Teams and constitutes the Core Programme (CP). The CP covered 35% of the total observing time during the first year and decreased to 30% in the second year, and to 25% in the third year. The scientific topics of the CP are focused on regular scans of the Galactic plane and deep exposures on the central radian of the Milky Way and other specific regions (Norma and Scutum Arms, Vela and Virgo regions). One year after the observation, all the data become public*.

2. Science

The main scientific topics addressed by *INTEGRAL* satellite observations can be divided into: *Compact objects:* neutron stars, anomalous X-ray pulsars, soft-gamma repeaters, black holes; *The Milky Way:* the Galactic Centre and Sgr A*, diffuse and line emission; *Stellar nucleosynthesis:* hydrostatic (AGB, Wolf-Rayet) and explosive (SN, Novae); *Extragalactic astrophysics:* active galactic nuclei (seyferts, blazars, radiogalaxies), clusters of galaxies, gamma-ray bursts (GRB); *Identification of high-energy sources:* EGRET unidentified sources, new discoveries; *Unexpected discoveries.*

*For more details, see the documentation enclosed to each Announcement of Opportunity available at `http://integral.esac.esa.int/`.

In the following we shortly present a selection of the most important results obtained with *INTEGRAL*, with apologies to those about which we cannot report, because of space limitations.

2.1. *The Galactic Centre*

One of the most important results was obtained thanks to the unprecedented angular resolution of IBIS/ISGRI. The problem was the understanding of the excess of diffuse emission in the Galactic centre: indeed, previous missions found a strong $\gamma-$ray diffuse emission toward the centre of the Milky Way, that was thought to be originated from the interaction of cosmic-rays with interstellar medium (see Pohl[16] for a review). However, this model cannot explain excesses in the spectrum around 100 keV and above 1 GeV. *INTEGRAL* performed a deep exposure (1.5 Ms) toward the central radian of the Galaxy, and demonstrated that the excess was due to point sources not resolved by previous experiments[17].

Again IBIS/ISGRI performed an accurate study of Sgr A* and found a new hard X-ray source IGR J1745.6 − 2901, consistent with the Galactic Centre within $\approx 1'$, so that the association with Sgr A* is not conclusive[18]. On the other hand, the flux of IGR J1745.6−2901 is not consistent with the diffuse and point sources emission within $10'$ extrapolated from *Chandra* and *XMM-Newton* observations. These two X-ray satellites have observed some transients in the ISGRI error box, but these are too weak and soft to be associated with the hard X-ray source.

Another newly discovered *INTEGRAL* source, namely IGR J17475 − 2822, has been associated with the giant molecular cloud Sgr B2[19]. To explain how a molecular cloud could emit hard X-rays, it has been suggested that the radiation could be the result of a past activity of Sgr A* ($\approx 300-400$ years ago, when it behaved as an active galactic nucleus), that was Compton scattered and reprocessed by Sgr B2[19].

The spectrometer SPI focused on the electron-positron annihilation line (511 keV) and found a significant ($\approx 50\sigma$) emission from the Galactic bulge[20]. This emission region is centered on the Galactic centre, is symmetric, and is about 8° wide. There is no evidence of the positive latitude enhancement found by OSSE[21] and the emission along the Galactic disk is constrained with low significance ($\approx 4\sigma$).

2.2. *Compact objects*

Thanks to its characteristics, *INTEGRAL* can give important contribution mainly in the astrophysics of compact objects. Indeed, with a sensitivity of about $(0.8 - 1) \times 10^{-10}$ erg cm^{-2} s^{-1} in the $20 - 40$ keV energy band ($10 -$ 15 mCrab) for a 3σ detection in 2 ks, the IBIS/ISGRI detector is efficiently used to monitor the X-ray binaries population in the Milky Way with the regular scans of the Core Programme. On December 2, 2004, during these scans, it was discovered the fastest millisecond pulsar ever known: it is the new source IGR J00291 + 5934, that has a pulsation period of 1.67 ms[22]. In December 2004, *INTEGRAL* was the first, of more than twenty satellites, to detect a gigantic flare from the soft γ−ray repeater SGR 1806 − 20, so intense to significantly ionize the Earth's upper atmosphere[24,25,26].

The high angular resolution of *INTEGRAL* in the hard X-rays results crucial to disentangle the emission of different sources in crowded fields: for example, Paizis et al.[23] succeeded in studying GX 5−1, that is apparently located close ($40'$) to the black hole GRS 1758 − 258. Another case is represented by 4U 1630 − 47, a black hole in the Norma Arm: Tomsick et al.[30] have shown that the extreme behaviour during the period 2002−2004 is not in agreement with the classical definitions of spectral states.

Finally, among black holes, Cyg X-1 has been extensively studied during the Performance Verification phase, as it was the selected target for the first light of the main instruments[27,28,29].

2.3. *New sources and survey*

On January 29, 2003 *INTEGRAL* discovered its first new source[31], that was called IGR J16318 − 4848. It is located in the Norma Arm of the Galaxy, an active star forming region, where several other sources have been found later[†]. The high energy spectrum, from soft to hard X-rays as result of a joint observation of *XMM-Newton* and *INTEGRAL*, together with optical data, suggest that IGR J16318 − 4848 could be an accreting neutron star with a high-mass companion, enshrouded by a Compton-thick environment ($N_{\rm H} \approx 10^{24}$ cm^{-2}), perhaps the progenitor of a new population of X-ray binaries[32]. See also Kuulkers[33] for a recent review on this type of sources.

Recently, one of the new *INTEGRAL* sources, IGR J18135 − 1751, has

[†]See the web page on the new *INTEGRAL* sources maintained by J. Rodriguez at `http://isdc.unige.ch/~rodrigue/html/igrsources.html`

been associated with an unidentified TeV source discovered by the HESS telescope[34], namely HESS J1813 − 178. The spectral energy distribution suggests that the object could be a pulsar wind nebula embedded in its supernova remnant[35]. Soon after, *INTEGRAL* discovered the soft γ−ray emission from another unidentified TeV source (HESS J1837 − 069), that appears to be another pulsar or supernova remnant[36].

Several other new sources have been discovered by *INTEGRAL* up to date: the first IBIS/ISGRI source catalog[37], referring to the data from February to October 2003, contains 123 known and 28 new sources down to a flux limit of $\approx 1.7 \times 10^{-11}$ erg cm^{-2} s^{-1} in the $20 - 100$ keV energy range (1 mCrab). The latest publicly available update back to a presentation by A. Bird[‡] in January 2005 and reported 229 known and 31 new sources.

2.4. *Gamma-Ray Bursts (GRB)*

INTEGRAL has an elongated orbit with 72 hours period. This, in addition to the availability of the two ground stations of Redu and Goldstone, allows a continuous monitoring of the data in near-real time. This is performed by the *INTEGRAL Burst Alert System* (IBAS[38]), in order to distribute the GRB coordinates to the astronomical community with unprecedented coupling of high accuracy ($\approx 3'$) and low time delay ($\approx 20 - 30$ s). See Mereghetti & Götz[39] for a review of the IBAS activity. Among the several GRB detected by *INTEGRAL*, it is worth mentioning the first that was in the field of view of the instruments (IBIS, SPI). It occurred on November 25, 2002, during the Performance Verification phase[40,41]. Moreover, *INTEGRAL* detected also the GRB nearest to the Earth ($z = 0.106$)[42].

Other efficient ways to detect GRB (particularly out of the instruments FOV) are available on board *INTEGRAL* satellite with the Anticoincidence Shield (ACS) of SPI[43], the IBIS Compton mode (that makes use of both layers ISGRI and PICsIT)[44], and the spectral timing mode of IBIS/PICsIT[45]. In these ways, it is possible to reach energies up to a few MeV.

2.5. *Active Galactic Nuclei*

Although AGN are numerically much less than other sources detected by *INTEGRAL*, there are some interesting studies worth mentioning. Six

[‡]Talk given at the *Internal INTEGRAL Science Workshop* held at ESA/ESTEC on 18 − 21 January 2005. See `http://integral.esac.esa.int/workshops/Jan2005/`.

blazars have been detected by *INTEGRAL* to date and it is worth noting that these are all also EGRET sources (i.e., that have been detected at energies greater than 100 MeV by the EGRET experiment on board the *Compton Gamma-Ray Observatory*). Three of these six AGN have been detected during a multiwavelength Target-of-Opportunity (ToO) programme, led by E. Pian, organized to study the blazars in outburst. These are: S5 $0716+714$[46], following an outburst started at the end of March 2004; 3C 454.3[47], recently observed during the long outburst of May 2005; and S5 $0836+710$, serendipitously detected in the field of view of the observation of S5 $0716+714$[46]. The remaining three blazars are the well known 3C 273[48] and 3C 279[49] in the Virgo region, and PKS $1830-211$[50], the farthest object ($z = 2.51$) detected to date by *INTEGRAL*.

Still in the field of view of the ToO on S5 $0716+714$ are present two more AGN: the Seyfert active nuclei Markarian 3 and Markarian 6[46]. The data from Mkn 3 show a hint for the presence of a break in the spectrum between 60 and 100 keV, but the low statistics did not allow to better constrain the value[46].

Among the Seyferts, other studies have been done on NGC 4388[51], joint with *XMM-Newton*, and GRS $1734-292$[52], a luminous Seyfert 1 behind the Galactic Centre. This latter AGN is also the only source detected by *INTEGRAL* inside the 99% probability contours of the EGRET unidentified source 3EG J1736$-$2908 and thus opening the possibility that the AGN could be the counterpart of the EGRET source[53]. However, since no other Seyferts are known to emit at $E > 100$ MeV, if this association will be confirmed by future observations (with *GLAST*, for example) it remains to understand what is the physical mechanism able to generate high-energy $\gamma-$rays from a Seyfert active nucleus.

Other EGRET AGN observed by *INTEGRAL* are the radio galaxies NGC 6251[54] and Centaurus A[55]. In the former case, a faint doubtful detection was strengthened by the agreement of the measured flux with the spectral energy distribution expected from a synchrotron self-Compton model.

Several other AGN, together with a detection in the $30-50$ keV energy band of the Coma Cluster, are present in the field of view analyzed by Krivonos et al.[56]. This allowed to evaluate the extragalactic hard X-ray background as observed by *INTEGRAL*, but given the small sky area covered (3%), much larger samples are needed to better asses the source contribution in this energy band.

3. Final Remarks

Thanks to its main characteristics in the hard X/soft-γ rays (high sensitivity, high angular resolution, high spectral resolution, wide FOV), at the end of the third year of operation *INTEGRAL* has significantly contributed to the advancement of the $\gamma-$ray astrophysics with several important contributions, a few of these have been shortly reviewed here. In addition, a public data archive is available to the astronomical community[§], where gigabytes of data are still waiting to be analyzed. We hope that this short review can be useful to show the huge work done to date, the work still to be done, and the potentialities of *INTEGRAL* in the exploration of the $\gamma-$ray sky.

References

1. Winkler C., Courvoisier T.J.-L., Di Cocco G., et al., 2003, A&A 411, L1
2. Jensen P.L., Clausen K., Cassi C., et al., 2003, A&A 411, L7
3. Much R., Barr P., Hansson L., et al., 2003, A&A 411, L49
4. Courvoisier T.J.-L., Walter R., Beckmann V., et al., 2003, A&A 411, L53
5. Ubertini P., Lebrun F., Di Cocco G., et al., 2003, A&A 411, L131
6. Vedrenne G., Roques J.-P., Schönfelder V., et al., 2003, A&A 411, L63
7. Lebrun F., Leray J.P., Lavocat P., et al., 2003, A&A 411, L141
8. Di Cocco G., Caroli E., Celesti E., et al., 2003, A&A 411, L189
9. Lund N., Budtz-Jørgensen C., Westergaard N.J., et al., 2003, A&A 411, L231
10. Mas-Hesse J.M., Giménez A., Culhane J.L., et al., 2003, A&A 411, L261
11. Goldwurm A., David P., Foschini L., et al., 2003, A&A 411, L223
12. Gros A., Goldwurm A., Cadolle-Bel M., et al., 2003, A&A 411, L179
13. Skinner G. & Connell P., 2003, A&A 411, L123
14. Strong A., 2003, A&A 411, L127
15. Westergaard N.J., Kretschmar P., Oxborrow C.A., et al., 2003, A&A 411, L257
16. Pohl M., 2002, Proceedings "The Universe viewed in gamma-rays", `astro-ph/0211622`
17. Lebrun F., Terrier R., Bazzano A., et al., 2004, Nature 428, 293
18. Bélanger G., Goldwurm A., Goldoni P., et al., 2004, ApJ 601, L163
19. Revnivtsev M.G., Churazov E.M., Sazonov S.Yu., et al., 2004, A&A 425, L49
20. Knödlseder J., Jean P., Lonjou V., et al., 2005, A&A, accepted for publication (`astro-ph/0506026`)
21. Purcell W.R., Cheng L.-X., Dixon D.D., et al., 1997, ApJ 491, 725
22. Shaw S., Mowlavi N., Rodriguez J., et al., 2005, A&A 432, L13

[§]Everything for the data analysis (data, software, documentation) can be found at the *INTEGRAL Science Data Centre* web page `http://isdc.unige.ch/`.

23. Paizis A., Ebisawa K., Tikkanen T., et al., 2005, A&A, accepted for publication (astro-ph/0507378)
24. Borokowski J., Götz D., Mereghetti S., et al., 2004, GCN 2920
25. Mereghetti S., Götz D., von Kienlin A., et al., 2005, ApJ 624, L105
26. Hurley K., Boggs S.E., Smith D.M., et al., 2005, Nature 434, 1098
27. Bouchet L., Jourdain E., Roques J.P., et al., 2003, A&A 411, L377
28. Pottschmidt K., Wilms J., Chernyakova M., et al., 2003, A&A 411, L383
29. Bazzano A., Bird A.J., Capitanio F., et al., 2003, A&A 411, L389
30. Tomsick J.A., Corbel S., Goldwurm A., Kaaret P., 2005, ApJ 630, 413
31. Courvoisier T.J.-L., Walter R., Rodriguez J., et al., 2003, IAU Circ., 8063
32. Walter R., Rodriguez J., Foschini L., et al., 2003, A&A 411, L427
33. Kuulkers E., 2005, Proceedings "Interacting Binaries: Accretion, Evolution and Outcomes", (astro-ph/0504625)
34. Aharonian F., Akhperjanian A.G., Aye K.-M., et al., 2005, Science 307, 1938
35. Ubertini P., Bassani L., Malizia A., et al., 2005, ApJ 629, L109
36. Malizia A., Bassani L., Stephen J.B., et al., 2005, ApJ 630, L157
37. Bird A.J., Barlow E.J., Bassani L., et al., 2004, ApJ 607, L33
38. Mereghetti S., Götz D., Borkowski J., et al., 2003, A&A 411, L291
39. Mereghetti S. & Götz D., 2005, Proceedings IV Workshop "Gamma-Ray Bursts in the Afterglow Era", (astro-ph/0506198)
40. Bazzano A. & Paizis A., 2002, GCN 1706
41. Malaguti G., Bazzano A., Beckmann V., et al., 2003, A&A 411, L307
42. Sazonov S.Yu., Lutovinov A.A., Sunyaev R.A., 2004, Nature 430, 646
43. Rau A., von Kienlin A., Hurley K., Lichti G.G., 2005, A&A 438, 1175
44. Marcinkowski R., Laurent Ph., Denis M., et al., 2005, "Localization & spectra of the Integral Compton mode gamma-ray bursts". "Astrophysical Sources of high-energy particles and radiation", 20-24 June 2005, Torun (Poland).
45. Marcinkowski R., Laurent Ph., Denis M., et al., 2005, "First catalogue of the Integral off-axis gamma-ray burts". "Astrophysical Sources of high-energy particles and radiation", 20-24 June 2005, Torun (Poland).
46. Pian E., Foschini L., Beckmann V., et al., 2005, A&A 429, 427
47. Foschini L., Di Cocco G., Malaguti G., et al., 2005, ATel 497
48. Courvoisier T.J.-L., Beckmann V., Bourban G., et al. 2003, A&A, 411, L343
49. Collmar W., Böttcher M., Burwitz V., et al., 2004, Proceedings V INTEGRAL Workshop "The INTEGRAL Universe", Editors: V. Schönfelder, G. Lichti & C. Winkler, ESA SP-552, p. 555 (astro-ph/0410302)
50. De Rosa A., Piro L., Tramacenere A., et al., 2005, A&A, 438, 121
51. Beckmann V., Gehrels N., Favre P., et al., 2004, ApJ 601, L163
52. Sazonov S.Yu., Revnivtsev M.G., Lutovinov A.A. et al., 2004, A&A 421, L21
53. Di Cocco G., Foschini L., Grandi P., et al., 2004, A&A 425, 89
54. Foschini L., Chiaberge M., Grandi P., et al., 2005, A&A 433, 515
55. Rothschild R.E., Wilms J., Benlloch S., et al., 2004, AAS HEAD Meeting 8, 04.01
56. Krivonos R., Vikhlinin A., Churazov E., et al., 2005, ApJ 625, 89

XMM OBSERVATIONS OF GEMINGA, PSR B1055-52 AND PSR B0656+14: PHASE RESOLVED SPECTROSCOPY AS A TOOL TO INVESTIGATE THE X-γ CONNECTION*

P. A. CARAVEO, A. DE LUCA, S. MEREGHETTI

INAF/IASF Milano
Via Bassini 15, I-20133 Milano, Italy
E-mail: pat@mi.iasf.cnr.it

G. F. BIGNAMI

Università degli Studi di Pavia,
Dipartimento di Fisica Nucleare e Teorica,
Via Bassi 6, I-27100 Pavia, Italy

CESR/CNRS-UPS
9, Avenue du Colonel Roche, 31028 Toulouse Cedex 4, France

XMM-Newton observations of PSR B0656+14, PSR B1055-52 and Geminga have increased dramatically the statistics available for these three isolated neutron stars (INSs) so that it is now possible to perform phase resolved spectroscopy for all three objects. The phase-averaged spectrum of the three INSs is best described by a three component model including two blackbody curves, a cooler one, possibly originating from the bulk of the star surface, and a hotter one, coming from a smaller portion of the star surface (a "hot spot"), plus a power law. Dividing the pulsars' light curves in 10 identical phase intervals, we can study for the first time the evolution of the spectral shape as a function of the stars' rotation. Very dramatic (and highly significant) changes are clearly detected at specific phases. The relative contributions of the three components are seen to vary as a function of the pulsars' phase, as the stars' rotation bring into view different emitting regions. The hot spots, which have very different apparent dimensions (in spite of the similarity of the three neutron stars polar cap radii) are responsible for the bulk of the phase variation. Thus, the rotation of the neutron stars' surfaces can be directly seen for the first time.

*Based on observations with XMM-Newton, an ESA science mission with instruments and contributions directly funded by ESA member states and the USA (NASA).

1. Introduction

PSR B0656+14, PSR B1055-52 and Geminga are nearby isolated neutron stars characterized by similar values of their periods and period derivatives. Thus, their derived ages (few 100 thousand years), their inferred magnetic fields (few 10^{12} G) and their energetics ($\dot{E}_{rot} \sim 10^{34}$ erg s^{-1}) are very similar. In X-rays they shine both owing to thermal and non-thermal mechanisms. Black-body emission, from the majority of their surface, is seen to be complemented at higher energies by power law "tails". They have optical counterparts of m_v ~26 (Bignami et al., 1987: Geminga; Caraveo et al., 1994: PSR B0656+14; Mignani et al., 1997: PSR B1055-52), compatible, or slightly brighter than the optical magnitudes expected on the basis of the simple extrapolation of their X-ray black body emissions. While PSR B1055-52 and Geminga are bona fide gamma-ray pulsars (Fierro et al., 1993; Bertsch et al., 1992), for PSR B0656+14 a tentative detection awaits confirmation (Ramanamurthy et al., 1996).

Recently, these three INSs have been observed by XMM-Newton which, collecting data both in imaging and timing mode, yielded an exceptional harvest of X-ray photons over the ample energy interval covered by the EPIC cameras. Here we shall exploit the XMM-Newton database to access the NSs spectral behaviour throughout their rotation.

2. The EPIC data

The EPIC data come after two decades of X-ray observations of isolated neutron stars, starting with their discovery as soft, probably thermal sources by the Einstein Observatory. For a recent review of the observational panorama on the three INSs, as well as for a discussion of the data analysis technique, see De Luca et al.(2005).

With a number of time-tagged photons more than doubling all previous statistics, and with a wider (0.15 to 8 keV) spectral grasp, EPIC offers now the first chance of meaningful phase-resolved spectroscopy for the three INSs.

Following the procedure outlined by Caraveo et al. (2004):

- first, we address the phase averaged spectral shape of the three objects to obtain useful pieces of information on the components needed to fit their overall spectra;
- next, we perform the timing analysis to search for the best period and to build their light curves;

- then, we divide the folded light-curves in 10 phase intervals and extract the spectra corresponding to each phase interval;
- finally, we address the variation of each spectrum wrt. the average one.

2.1. *Phase-integrated spectral analysis*

As in the case of Geminga (Caraveo et al., 2004), the best fitting model is found combining two blackbody curves and a power law, confirming the results reported by Pavlov et al. (2002) (for both PSR B0656+14 and PSR B1055-52) and Becker & Aschenbach (2002) (for PSR B1055-52) on the basis of a preliminary analysis of Chandra and XMM-Newton data, respectively. Details on the spectral fits may be found in De Luca et al.(2005). We note that a single-temperature, magnetized neutron star atmosphere model (Zavlin et al. 1996) cannot adequately reproduce the profile of the low-energy part of the spectrum of PSR B0656+14 and PSR B1055-52, similarly to the case of a single blackbody curve. Moreover, such model requires a very large emitting surface (with emitting radii of $\sim$100 km).

The lower temperature blackbody (hereafter "cool blackbody") is associated to an emitting region with an area compatible with the full surface of the neutron star. The higher temperature component ("hot blackbody") is seen to originate from a much smaller area.

Although the overall shape of the spectrum of the three INSs is very similar, remarkable differences in the luminosity of the different spectral components are immediately evident. In particular, we note that the luminosity of the hot blackbody in the case of Geminga is $\sim$ 2 orders of magnitude lower than in the other two cases.

We will now exploit the unprecedented harvest of photons to perform phase-resolved spectroscopy of the three neutron stars.

2.2. *Timing analysis*

As a first step, we have studied the high time resolution pn data in order to compute the period of our targets. Source photons were selected in the energy range 0.15-8 keV for pn Small Window mode and 0.15-2 keV for pn Fast Timing mode. Photon time of arrivals were converted to the Solar System barycenter using the SAS task *barycen* and folded using 10 bins in a range of trial periods around the expected values. A very significant detection of the pulsation was obtained in each case.

The measured period values are in perfect agreement with our expecta-

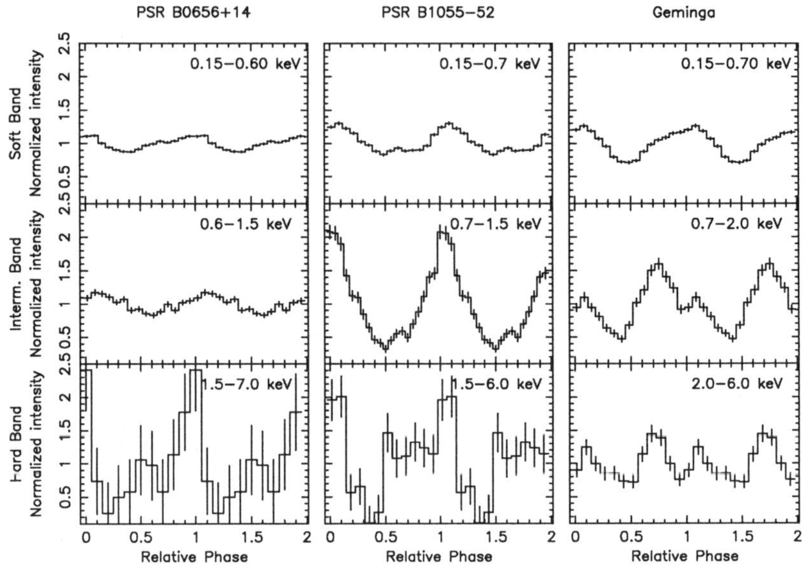

Figure 1. Energy-resolved light curves of PSR B0656+14, PSR B1055-52 and Geminga in different energy ranges. To ease the comparison of the behaviour of the three INSs, all light curves have been plotted setting phase 0 to the X-ray maximum. Pulsed fractions (computed as the ratio between the counts above the minimum and the total number of counts) are as follows: PSR B0656+14 12.3±0.4% in 0.15-0.6 keV, 16.9±2.3% in 0.6-1.5 keV, 75±20% in 1.5-7.0 keV; PSR B1055-52 16.7±0.6% in 0.15-0.7 keV, 67±3% in 0.7-1.5 keV, 90±10% in 1.5-6.0 keV; Geminga 28.4±0.6% in 0.15-0.7 keV, 54.5±2.4% in 0.7-2.0 keV, 33±5% in 2.0-6.0 keV.

tions on the basis of radio ephemeris (for PSR 1055-52 and PSR0656+14) and the gamma-ray data in the case of Geminga .

Figure 1 shows the pulsars' pulse profiles in three different energy intervals. The use of normalized counts makes it easy to see the different pulsed fractions characterizing the three objects as well as their variation in the different energy ranges. Note that the energy ranges used are different for different targets, since for each source the energy intervals have been chosen in order to visualize the light curve relative to the cool blackbody, hot blackbody and power law.

2.3. *Phase-resolved spectroscopy*

To assess the spectral behaviour throughout the pulsars' rotational phase we have divided their light curves in 10 equal phase intervals We then extracted 10 phase-resolved spectra. Each spectrum was rebinned in order to have at least 40 counts per bin.

As already suggested by the analysis of the energy-resolved pulse profiles, the spectra of the three sources do vary as a function of the pulsars' rotational phase. This is easily seen in the left panels of Fig.2,3,4, where the 10 phase-resolved spectra are superimposed (an animated version of the 10 phase resolved spectra of Geminga, PSR B1055-52 and PSRB 0656+14 is available at http://www.mi.iasf.cnr.it/~deluca/3musk/). The variation is dramatic in the case of PSR B1055-52 and Geminga, owing to their high phase modulation, while it is less spectacular (but still seen with high significance) for PSR B0656+14, due to its lower modulation.

3. Analysis of the phase resolved spectra

Indeed, the X-ray emission from our three targets is a mixture between surface thermal emission and magnetospheric, non-thermal one. While neutron star physics is reflected in their photon spectra, geometrical constraints, such as viewing angles of rotational and magnetic axes, shape their observed light curve. Indeed, photons emitted by pulsars carry the signature of their production mechanisms as well as of the geometry of their emitting regions. Phase modulation takes place as different emitting regions are brought into view during the star rotation. Geometry can also influence source spectral shapes, because of different emission mechanisms in different regions. Thus, the spectral variability unveiled by XMM-Newton for Geminga, PSR B0656+16 and PSR B1055-52 as a function of the pulsars' rotation can be used to assess how the neutron stars' emitting surfaces do change as they rotate. Following Caraveo et al.(2004), we have used the phase-integrated best fit model as a template to describe the phase-resolved data. The N_H value, the temperatures of both blackbodies and the power law photon index were fixed to the values best fitting the phase-integrated spectrum. The phase modulation was then reproduced by allowing the normalization parameters of each spectral component to vary independently.

Such a simple approach (using only three free parameters) yielded a satisfactory description of the phase-resolved spectra for Geminga, PSR B0656+14 and PSR B1055-52. Reduced χ^2 of $\sim 0.8 - 1.4$ were obtained in different phase intervals.

The results of phase-resolved spectral fits are described in De Luca et al.(2005) (an animated view of the phase-resolved fits can be found in http://www.mi.iasf.cnr.it/~deluca/3musk). Knowing the objects' distance values, we can compute the emitting surface corresponding to each blackbody curve. To visualize the evolution of the spectral parameters as a function of the pulsars' phase, we have plotted the blackbody radii values, as well as the power law intensities, as a function of the objects' rotational phase. This is shown in the right panels of Fig.2, 3, 4 for PSR B0656+14, PSR B1055-52 and Geminga, respectively.

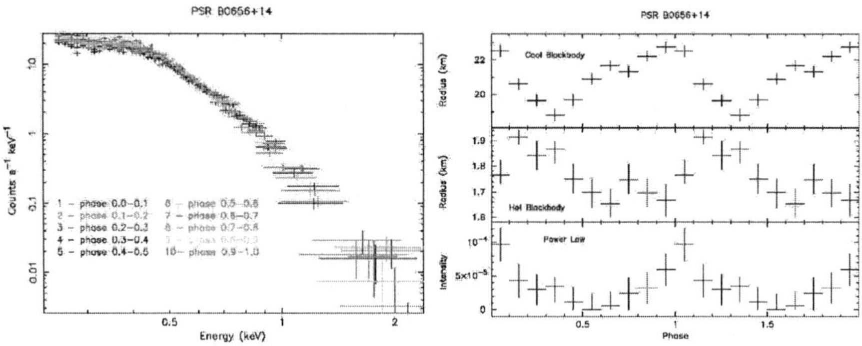

Figure 2. Left panel: observed phase-resolved spectra for PSR B0656+14. Right panel: phase-resolved spectral parameters plotted as a function of the pulsar phase.

The behaviour of the three neutron stars as a function of their rotational phase is vastly different.

- The cool blackbody component shows a similar phase evolution for the three targets. The values of the blackbody radii follow a roughly sinusoidal profile with a modulation $\leq$10% (wrt. the average value) for PSR B0656+14 and for PSR B1055-52, of $\sim$15% for Geminga.
- The hot blackbody components have a similar phase evolution with a sinusoidal profile. Hovever, we see striking differences both in the amplitude of their phase modulation and on their overall luminosity. While for PSR B0656+14 the modulation in the emitting radius wrt. the average value is <10%, similar to the value found for the cool blackbody component, in the case of PSR B1055-52 we see a 100% modulation, since the hot blackbody component is not seen in 4 out of 10 phase intervals. A similar, 100% modulation is observed also in the case of Geminga, although in this case the hot blackbody component is seen to disappear in just one phase interval, and the profile of its phase evolution is markedly broader.

As far as the radii values are concerned, we note that they span a factor of 30, ranging from the 60 m of Geminga to the ~2 km of PSR B0656+14.

- The relative phase evolution of the three spectral components is different for the three neutron stars. This is particularly evident when looking at the two blackbody components. In the case of PSR B0656+14 some sort of anti-correlation is observed, with the hot blackbody peaking in correspondence of the cool blackbody minimum. This is at odds with the behaviour observed for PSR B1055-52, where the two thermal components appear definitely correlated. Geminga shows an intermediate phenomenology: the cool and the hot blackbodies have a ~ 0.25 difference in their phase pattern.

4. Discussion

For all objects the hot blackbody is the most dramatically variable spectral component. It is natural to interpret such marked variations as an effect of the star rotation, which brings into view and hides one or more hot spots on the star surface. It is commonly accepted that neutron stars should have polar caps hotter than the rest of the surface. This could be due to different processes such as the bombardment of charged particles accelerated in the magnetosphere and falling back to the polar caps along magnetic field lines (return currents, see e.g. Ruderman & Sutherland 1975; Arons & Scharlemann 1979), or anisotropic heat transfer from the neutron star core, which depends strongly on the magnetic field direction and is maximum along the magnetic field lines (see e.g. Greenstein and Hartke 1983). An association of the observed rotating hot spot(s) with the neutron stars' polar caps seems therefore rather obvious. Straight estimates of neutron star polar cap size, based on a simple "centered" dipole magnetic field geometry (polar cap radius $R_{PC} = R\sqrt{\frac{R\Omega}{c}}$), where R is the neutron star radius, Ω is the angular frequency and c is the speed of light), predict very similar radii for the three neutron stars, characterized by similar periods (233 m for PSR B0656+14, 326 m for PSR B1055-52 and 297 m for Geminga, assuming a standard neutron star radius of 10 km). The observed radii are instead markedly different, with values ranging from ~ 60 m for Geminga to ~2 km for PSR B0656+14. Psaltis et al.(2000) showed that the observed polar cap radius may be different from the actual one by a large amount, due to geometrical effects. Indeed, Caraveo et al.(2004) proposed a highly inclined viewing angle to reduce the surface of the emitting region in the case of Geminga. PSR B0656+14 and PSR B1055-52 face a completely different situation, since their polar caps are significantly larger than expected. Standard estimates, based on pure geometrical considerations, are clearly unsatisfactory. Even using the unrealistically large neutron star radius found for PSR B0656+14, the inferred polar cap dimension is <600 m, far less than the hot spot radius. It is also possible, as suggested by Ruderman (2003), that thermal photons be reprocessed higher up in the magnetosphere, interacting with charged particles. Such mechanism would obviously bias our measurements of emitting radii based on blackbody fits.

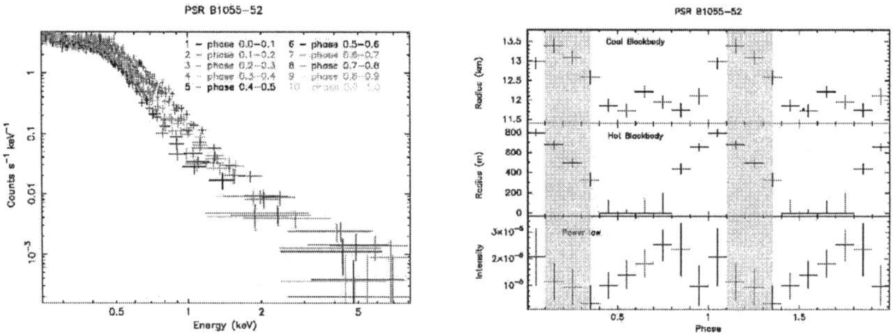

Figure 3. Same as Fig.2 for the case of PSR B1055-52. The shadowed region in the right panel marks the ~ 0.2 phase interval corresponding to the γ-ray peak(s) observed by EGRET.

The observed hot spot bolometric luminosities vary by more than 2 orders of magnitude, from the rather similar values found for PSR B0656+14 and PSR B1055-52 (5.7×10^{31} erg s^{-1} and 1.6×10^{31} erg s^{-1}, respectively), to the much dimmer Geminga (1.6×10^{29} erg s^{-1}). The former two cases are in broad agreement with the theoretical expectations of Harding & Muslimov (2002) for polar cap heating due to downstreaming of $e^{+/-}$ generated by curvature radiation photons. In the case of the Geminga pulsar, close to the death line for creation of $e^{+/-}$ couples via curvature radiation, the lower polar cap luminosity is consistent with expectations for polar cap heating due to bombardment by particles created by inverse Compton scattered photons only (Harding & Muslimov, 2002; see Caraveo et al., 2004).

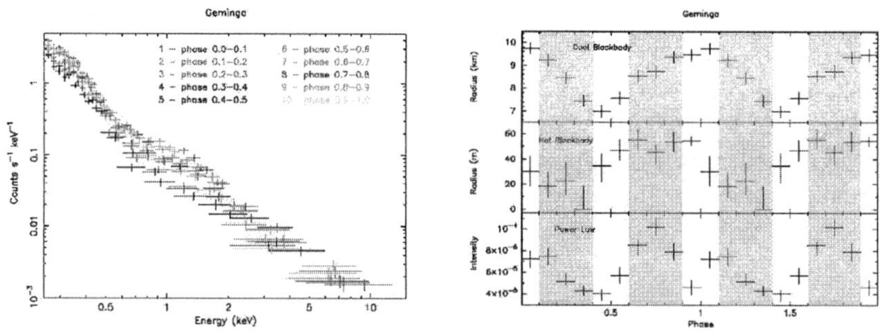

Figure 4. Same as Fig.2 and Fig.3 for the case of Geminga. The shadowed regions mark the phase intervals corresponding to the γ-ray peaks observed by EGRET. The highest γ-ray peak occurs at phase 0.25 ± 0.15, the second one at phase 0.75 ± 0.15 (the uncertainty is due to the extrapolation of the EGRET ephemeris to the epoch of the XMM-Newton observation).

5. Conclusions

In spite of the potential interest of phase-resolved spectroscopy as a tool to probe the changing environment of a rotating neutron star, the paucity of the number of X-ray photons so far available made it impossible to routinely apply this method. EPIC on XMM-Newton and Chandra have changed our scenario. High throughput and long observations can now provide an adequate harvest of time-tagged photons, perfectly suited for the task. Long EPIC observations of Geminga, PSR B0656+14 and PSR B1055-52 yielded an unprecedented combination of abundant statistics and spectral grasp, thus rendering possible, for the first time, detailed phase-resolved spectroscopy of their pulsed emission.

In general, the new X-ray phenomenology revealed by the EPIC/XMM-Newton combination presents new aspects of these three isolated, nearby neutron stars. Rotating, polar hot spots are clearly detected for the first time. It is tempting to link their origin to energetic particle bombardment, following the Ruderman mechanism, especially because Geminga and PSR B1055-52 (and possibly PSR B0656+14) are strong γ-ray sources. In the case of Geminga, moreover, Caraveo et al. (2003) have found a truly "smoking gun" evidence for this: the object's bow shock is filled with electrons of just the right energy ($\sim 10^{14}$ eV) and luminosity (a few 10^{28} erg s^{-1}) to be the particles escaping from a polar cap.

However, the measurement of polar cap sizes, rendered accurate by the excellent distance measurements, does not match simple theoretical expectations. 3-D geometry as well as beaming physics will have to be invoked for detailed models. The same holds true for the apparent sizes of the cool blackbody components. At least in one case they largely exceed the size of any decent neutron star. Simple atmospheric models only make matters worse.

Finally, the apparent puzzle posed by the difference in component phase alignements for the three objects is difficult to reconcile with existing pictures.

Acknowledgments

The XMM-Newton data analysis is supported by the Italian Space Agency (ASI). ADL acknowledges an ASI fellowship.

References

1. Arons, J. & Scharlemann, E.T., 1979, ApJ 231, 854
2. Becker, W. & Aschenbach, B.M., 2002, Proceedings of the 270. WE-Heraeus Seminar on Neutron Stars, Pulsars, and Supernova Remnants. Edited by W. Becker, H. Lesch, and J. Trümper. Garching bei München: Max-Plank-Institut für extraterrestrische Physik, p.64
3. Bertsch et al., 1992, Nature 357, 306
4. Bignami, G.F., Caraveo, P.A., Paul, J.A., Salotti, L., Vigroux, L., 1987, ApJ 319, 358
5. Caraveo, P.A., et al., 2003, Science 301, 1345
6. Caraveo, P.A., De Luca, A., Mereghetti, S., Pellizzoni, A., Bignami, G.F., 2004, Science, 305, 376
7. Caraveo, P.A., Bignami, G.F., Mereghetti, S., 1994 ApJ 422, L87
8. De Luca, A., Caraveo, P.A., Mereghetti, S., Negroni, M., Bignami, G.F., 2004, ApJ 623, 1051
9. Fierro, J.M., et al., 1993, ApJ 413, L27
10. Greenstein, G. & Hartke, G.J., 1983, ApJ 271, 283
11. Harding, A.K. & Muslimov, A.G., 2002, ApJ 568, 862
12. Mignani, R.P., Caraveo, P.A., Bignami, G.F., 1997, ApJ 474, L51
13. Pavlov, G.G., Zavlin, V.E., Sanwal, D., 2002, Proceedings of the 270. WE-Heraeus Seminar on Neutron Stars, Pulsars, and Supernova Remnants. Edited by W. Becker, H. Lesch, and J. Trümper. Garching bei München: Max-Plank-Institut für extraterrestrische Physik, p.273
14. Psaltis, D., Oezel, F., DeDeo, S., 2000, ApJ 544, 390
15. Ramanamurthy, P.V., Fichtel, C.E., Kniffen, D.A., Sreekumar, P., Thompson, D.J., 1996, ApJ 458, 755
16. Ruderman, M. & Sutherland, P.G., 1975, ApJ 196, 51
17. Ruderman, M., 2003, 4th AGILE Science Workshop, "X-ray and Gamma-ray Astrophysics of Galactic Sources", astro-ph/0310777
18. Zavlin, V.E., Pavlov, G.G., Shibanov, Y.A., 1996, A&A 315, 141

Part IV

POSTER SESSION

SOFTWARE TIME-CALIBRATION OF THE ARGO-YBJ DETECTOR

A.K. CALABRESE MELCARNE
(for the ARGO-YBJ Collaboration)
INFN and Dipartimento di Fisica dell'Università di Lecce,
73100 Italy
E-mail: melcarne@le.infn.it

A software procedure for the time-calibration of the detector ARGO-YBJ has been implemented and applied to a portion of the apparatus presently in data taking.

1. Introduction

The ARGO-YBJ detector is a full coverage carpet made by Resistive Plate Chambers (RPCs) devoted to cosmic ray studies and γ-astronomy. Currently a portion of $\sim 1900\ m^2$ is instrumented and takes data since 2004. The RPCs are equipped with pick-up strips and the fast-OR signal of 8 strips constitutes the logical pixel (called pad) for triggering and timing purposes. In this way the detector is able to measure the arrival direction of the primary cosmic rays by detailed space-time pictures of the air showers. The front of the incoming particles is parameterized, in a first approximation, by a plane and the arrival times of the particles are fitted by minimizing the following quantity:

$$\chi^2 = \sum_i w_i \left(t_i - t_0 - \frac{l}{c} x_i - \frac{m}{c} y_i \right)^2 \qquad (1)$$

where the sum is over the fired pads, w_i is the number of strips fired in the i-th pad, t_i is the measured time and x_i,y_i are the pad coordinates. The parameters reconstructed by the fit are the direction cosines l, m and the time constant t_0. The accuracy in the reconstruction of the shower direction depends on the measurement of the arrival times of the particles on the pads.

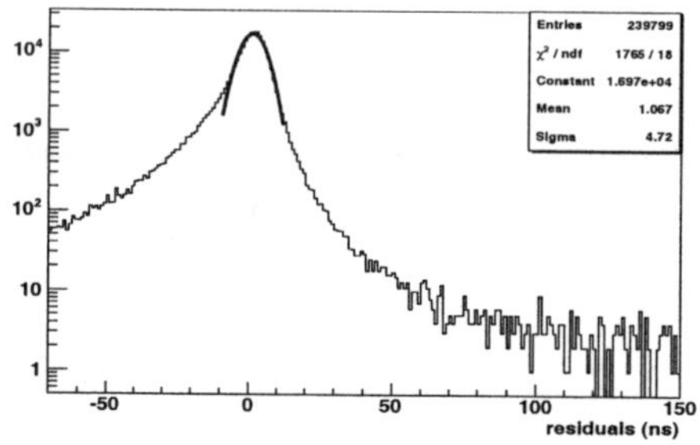

Figure 1. Residual distribution for a pad. The gaussian fit is suprimposed on the peak

2. Calibration procedure

The calibration has been performed on a sample of $\sim 6.5 \times 10^{+6}$ events collected in January 2005 with a trigger requiring at least 60 fired pads. The first step of the procedure is to reduce the pad residuals, i.e. the difference between the measured time and the fit-time as in Eq. (1). The correction is estimated fitting the peak of the residual distribution for each pad, as shown in Fig. 1. The mean value of the Gaussian fit is the pad-residual. It is used to correct the times measured by the pad. The distribution for all the pad-residuals before the calibration has a range of ~ 25 ns (Fig. 2).

The residual correction improves the quality of the shower reconstruction: the mean value of χ^2 decrease from 121.3 ns^2 to 94.2 ns^2, the new pad-residual distribution (Fig. 2) is narrower than the previuos one and the anomalous structure at higher values disappeares. Moreover the mean

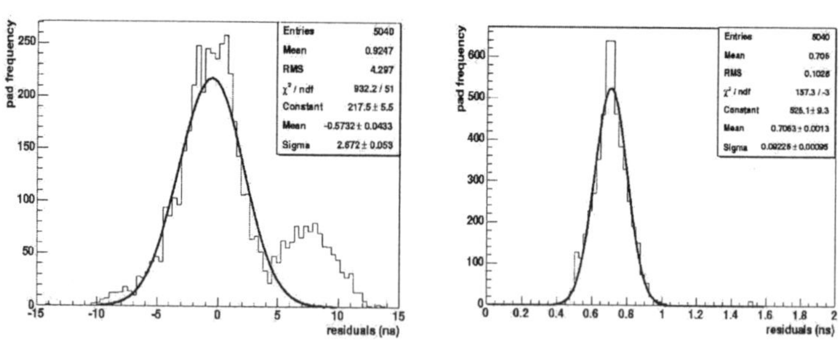

Figure 2. Left panel shows the pad-residual distribution before the calibration, while the right one shows the pad-residual distribution after the calibration. The gaussian fit is superimposed.

value of the reconstructed zenith angle (θ) decreases as expected [1]. But these improvements do not demonstrate that the absolute pointing is correct. The distribution of the time offsets of the pads defines an average plane (in the space x, y and t) for the reconstruction of the showers. The orientation of this plane is casual and it is not modified by the residual correction. If we expect that the primary flux is symmetrical with respect to the azimuth angle (see Sec. 3), in first approximation, the mean values of the direction cosines have to be null and the azimuth distribution has to be uniform. So a second correction is needed that is a planar time-correction [1,2,3] that removes completely the asimmetry in the ϕ-distribution. The characteristic-plane correction will not modify the pad-residual distribution. These corrections become out exploiting the additive character of direction cosines:

$$\langle l \rangle = \langle l_0 \rangle + a = a \qquad \langle m \rangle = \langle m_0 \rangle + b = b \tag{2}$$

where l_0, m_0 are the true direction cosines of the showers, and $\langle l \rangle$, $\langle m \rangle$ are the mean reconstructed direction cosines to be used to correct the times. The total time-correction for i-th pad results

$$t_i \rightarrow t_i - \delta_i - \frac{a}{c}x_i - \frac{b}{c}y_i \tag{3}$$

where the residual correction is δ_i and the plane correction depends on a,b and on the pad coordinates x_i, y_i. The Fig. 3 on the left shows the TDC mean values with respect to the position of the pads after the calibration.

A circular symmetry is present as foreseen, the shape of the azimuth distribution depends on the trigger requirement and the TDC values are in

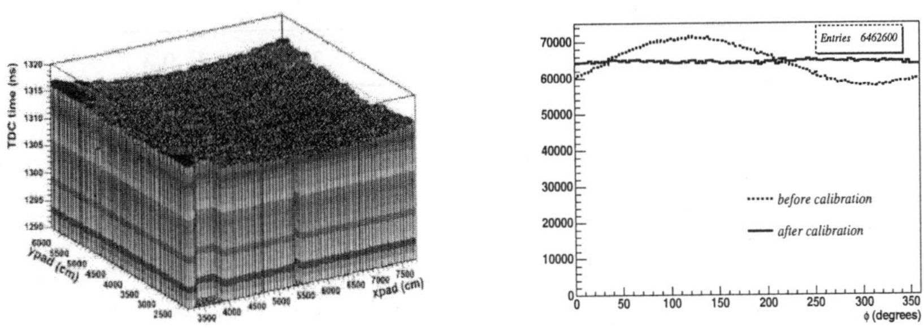

Figure 3. The left figure shows the TDC peaks versus the pad coordinates after the calibration while the right one is the azimuth distribution before and after the calibration procedure.

a range of few ns. The azimuth distribution is shown before and after the calibration in Fig. 3 on the right.

3. Study about the shape of the ϕ-distribution

The assumption of a flat azimuthal distribution is a good starting point but it is necessary to check the possibility of distortions due to physical or instrumental (trigger, detector geometry) effects. A preliminary Corsika simulation has been performed in order to study the effect of the geomagnetic field ($\theta_H \simeq 46°$ at the YBJ site) on the secondary particles in the shower [4,5]. The ϕ-distribution for real data before the calibration and for simulated data have been fitted with a two-harmonics function:

$$\frac{dN}{d\phi} = K[1 + A_I cos(\phi - \psi_I) + A_{II} cos(2\phi - \psi_{II})] \tag{4}$$

In Table 1. the modulation amplitudes are compared for simulated and real data. The values of A_I and A_{II} from real data are much higher as a confirm that the effect of a tilted characteristic plane masks the geomagnetic one. Then we can conclude that is necessary to apply a characteristic-plane correction.

Table 1. Harmonics amplitudes as a result of ϕ-distribution fit for for simulated and real data before calibration.

θ-range	A_I^{simu} (%)	A_{II}^{simu} (%)	A_I^{real} (%)	A_{II}^{real} (%)
$0° - 10°$	0.23	0.15	1.0	0.5
$10° - 20°$	1.2	0.25	3.4	0.2
$20° - 30°$	2.0	0.30	6.8	0.2
$30° - 40°$	2.8	0.28	13.8	0.4
$0° - 40°$	1.4	0.22	7.0	0.3

4. Calibration checks

A simple geometrical simulation permitted to verify that the sinusoidal shape is the effect of a tilted characteristic plane and the direction cosines of this plane are those expected from the $< l >$ and $< m >$ values and used for the time-correction. Finally, the stability in time of the calibration has been checked: after 3 months of data-taking the pad residuals results stable within 0.3 ns.

References

1. A.M. Elø and H.Arvela, 26th ICRC, Salt Lake City **5**, 328 (1999).
2. A.M. Elø and H.Arvela, 26th ICRC, Salt Lake City **5**, 324 (1999).
3. A.M. Elø and H.Arvela, 26th ICRC, Salt Lake City **5**, 320 (1999).
4. A.A. Ivanov , *JEPT Letters* **69**, 288 (1999).
5. M. Bahmanabadi et al., 27th ICRC, Hamburg 586 (2001), HE.128.

GAMMA-RAY BURST PHYSICS WITH GLAST

N. OMODEI

INFN sez. Pisa,
Edificio C - Polo Fibonacci -
Largo B. Pontecorvo, 3 PISA,
E-mail: nicola.omodei@pi.infn.it

The Gamma-ray Large Area Space Telescope (GLAST) is an international space mission that will study the cosmos in the energy range 10 keV-300 GeV, the upper end of which is one of the last poorly observed region of the celestial electromagnetic spectrum. The ancestor of the GLAST/LAT was the Energetic Gamma Ray Experiment Telescope (EGRET) detector, which flew onboard the Compton Gamma Ray Observatory (CGRO). The amount of information and the step forward that the high energy astrophysics made thanks to its 9 years of observations are impressive. Nevertheless, EGRET uncovered the tip of the iceberg, raising many questions, and it is in the light of EGRET's results that the great potential of the next generation gamma-ray telescope can be appreciated. GLAST will have an imaging gamma-ray telescope, the Large Area Telescope (LAT) vastly more capable than instruments flown previously, as well as a secondary instrument, the GLAST Bursts Monitor, or GBM, to augment the study of gamma-ray bursts. Gamma-Ray Bursts (GRBs) science is one of the most exciting challenges for the GLAST mission, exploring the high energy emission of one of the most intense phenomena in the sky, shading light on various problems: from the acceleration of particles to the emission processes, to more exotic physics like Quantum Gravity effect. In this paper we report the work done so far in the simulation development as well as the study of the LAT sensitivity to GRB.

1. The Gamma-ray Large Area Space Telescope (GLAST)

The main instrument on board the GLAST satellite is the Large Area Telescope (LAT). It is a pair conversion telescope, like EGRET, but the detectors will be based on solid-state technology, obviating the need for consumables and greatly decreasing instrument dead-time. The LAT comprises an array of 16 identical "tower" modules (see Fig. 1), each with a tracker (Si strips), a calorimeter (CsI with PIN diode readout) and a Data Acquisition (DAQ) module. The towers are surrounded by a finely segmented anti-coincidence detector (ACD), made by plastic scintillator, while the support structure is an aluminum strong-back "Grid" with heat pipes for transport of heat to the instrument sides.

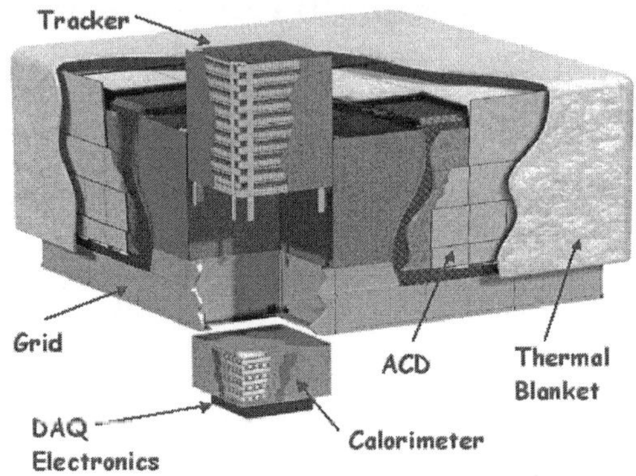

Figure 1. The LAT instrument components: The ACD is the anti coincidence detector, for the background rejection. CAL is the calorimeter which provide measurements on the energy. The Tracker is the complex tracker system based in Si Strips Detectors. The Grid has the structural function of hinging the towers. The DAQ electronic is mounted below. The thermal blanket covers the full instrument providing heat insulation.

LAT will have superior area, angular resolution, field of view, and dead time that together will provide a factor of 40 or more advance in sensitivity, covering the energy range between 20 MeV and 300 GeV. Aside from the main instrument LAT, a secondary instrument, the GLAST Burst Monitor (GBM), is foreseen. With the LAT and GBM, GLAST will be a flexible observatory for investigating the great range of astrophysical phenomena best studied in high energy gamma rays. NASA plans to launch GLAST in early 2007 from Kennedy Space Flight Center.

2. Gamma-Ray Bursts

Gamma-Ray Bursts (GRBs), are short and intense pulses of high energy radiation whose duration varies from fractions of a second to several hundred of seconds. They were serendipitously discovered in the sixties by the four Vela satellites[1] and their extragalactic origin were directly observed by the Italian-Dutch satellite BeppoSAX (1996-2002) measuring the first X-Ray afterglow[2]. BATSE (Burst And Transient Source Experiment) detector on board the CGRO observing more than 2700 burst, in its 9 years of observations while EGRET, mainly due to its small effective area, observed only few photons coming from GRBs[3], and the high energy emission from GRB is still puzzleing and enigmatic[4,5,6].

Several Gamma-Ray Bursts source models have been developed, starting from different hypothesis and describing different scenarios. One of the most creditable model is the *fireball* model introduced by Piran, 1999[7], starting from the evolution of a relativistic expanding shell[8]. In the most schematic view of the fireball model, a hidden central engine emits "shells" of matter (plasma) into the interstellar medium with relativistic bulk Lorentz factors (for a complete review of the Fireball model, see also[9]). If the central engine emits shells with different velocities, a faster shell can reach a slower one producing a shock. The dissipated energy can be used both to accelerate particles and to generate magnetic fields. Charged particles in magnetic fields lose energy via synchrotron emission and, eventually, can boost the synchrotron photons via inverse Compton (IC) scattering producing high energy photons. We have developed a GRB source model (details in[10]) based on the Fireball model, that can be used within the GLAST/LAT simulation framework[11] for studying the response of the LAT detector to GRB signal. In practice, we normalize GRB fluences to the BATSE observed distribution[12], and we extrapolate to LAT energies on the basis of common physical principle such as the synchrotron radiative process and the inverse Compton scattering. Here we are interested in two main features: the synchrotron cut-off due to the finite value of the Lorentz factor at which electrons are accelerated[13,14], and the inverse Compton component, in particular, how the sensitivity of the LAT instrument varies with respect to the inverse Compton emissivity.

2.1. *Measurement of the high energy cut-off*

In our simulation we assume as initial distribution of electrons a power law between γ_{min} and γ_{max} with power law index $-p$, and that each shock produced a pulse. The maximum Lorentz factor of the accelerated electrons (γ_{max}) determines a cut-off at high energies, so that we adopt a parameterized function for the GRB spectrum due to pure synchrotron emission. For fast cooling regime, which is the case of GRB: $P(E) = Exp(-E/E_M)(E/E_c)^{1/3}$ for energy less than the typical cooling energy (E_c); $P(E) = Exp(-E/E_M)(E/E_c)^{-1/2}$ for energies between the cooling energy and the "minimum" energy $E_c < E < E_m$; and $P(E) = Exp(-E/E_M)(E_m/E_c)^{-1/2}(E/E_m)^{-p/2}$ for energies higher that the minimum energy. Notice that for a γ_{max} large enough (as consequence, large E_M) the spectra reduces to the usual broken power law proposed by Sari et al.(1998)[15]. Nevertheless, there is a linear relation between the cut-off energy E_M, expressed in GeV, and the Lorentz factor of the emitting shell Γ, given, numerically by $E_M = 2.5\ \Gamma/100$. Measuring the high energy cut-off at GeV energies one can measure the (average) Lorentz factor of the expanding shells. In Fig. 2 we simulated a GRB with a BATSE fluence $3.17 \times 10^{-5} erg/cm^2$, which generates 1220 photons above 100 MeV, whose result in 408 triggered and reconstructed photons. The data have been fitted with a power law, and with a power law with an exponential cut-off. The chi-square test gives the probability for the different models, showing that the simple power law function is clearly unsatisfactory, while the power law with the exponential cut-off is in agreement with the (simulated) data. The cut-off estimated energy is 5.5 ± 1.5 GeV.

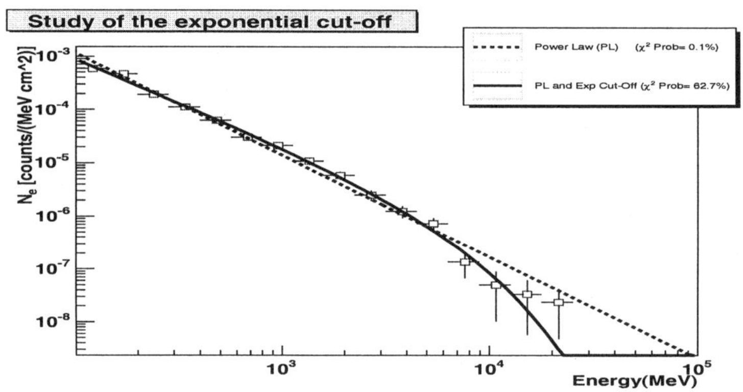

Figure 2. Study of the high energy cut-off. The simulated data have been produced using the GRB model forcing a high energy cut-off at 4.5 GeV. The reconstructed data are displayed in the plot and have been fitted with a power law distribution (dashed line), and with a power law distribution with an exponential cut-off (solid line). In the legend the values of the χ^2 test denote the probability to obtain a greater χ^2 by chance, for the two different models. The cut-off estimated energy is 5.5 ± 1.5 GeV.

2.2. *The inverse Compton component*

The inverse Compton process boosts up the synchrotron photons by means of scattering against the high energy electrons in "Self Synchrotron Compton" or SSC configuration. The electron distribution is a power law and the most probable scattering is between the lowest energetic electrons (whose energy is $\approx E_m$) against the synchrotron spectrum. In the relativistic case (when the energy of the electrons is much more greater than their rest mass) the photons are up scattered by a quantity γ^2 where the γ is the Lorentz factor of the electrons. In GRB, the peak of the synchrotron spectrum (E_{peak}) is typically of the order of hundreds of keV, and the minimum Lorentz factor of the accelerated particles is $\gamma \approx 10^2$: the typical IC peak is hence expected to be around GeV energies. The typical signature of the IC is the presence of a maximum in the $\nu F(\nu)$ spectrum (or $E^2 N(E)$) in the GeV energy range. We scale the inverse Compton component by means of a parameter τ which represents the intensity of the peak of the inverse Compton component (the height of the IC bump of the $\nu F(\nu)$ spectrum) relative to the intensity of the synchrotron component (the height of the synchrotron bump of the $\nu F(\nu)$ spectrum). Fig. 3 shows the simulation in case of $\tau = 1$, which represents, roughly, the same amount of energy in the synchrotron emissions as in the SSC emission.

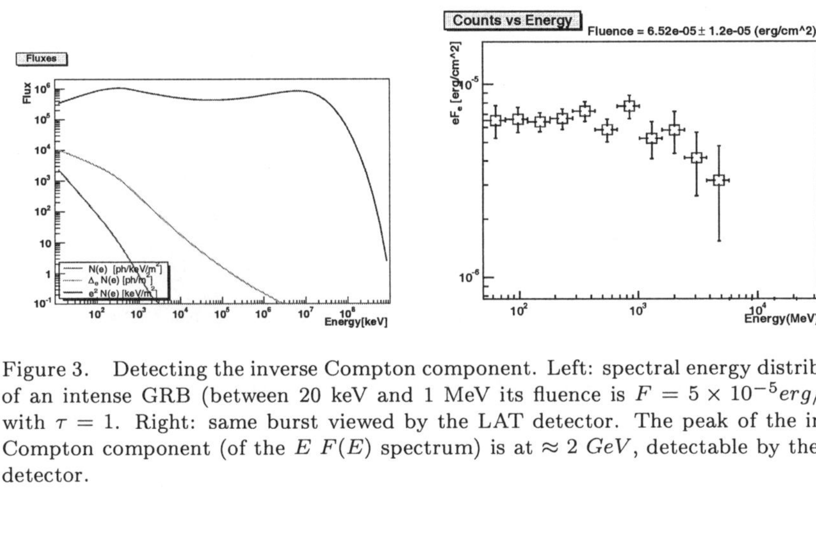

Figure 3. Detecting the inverse Compton component. Left: spectral energy distribution of an intense GRB (between 20 keV and 1 MeV its fluence is $F = 5 \times 10^{-5} erg/cm^2$) with $\tau = 1$. Right: same burst viewed by the LAT detector. The peak of the inverse Compton component (of the $E\ F(E)$ spectrum) is at $\approx 2\ GeV$, detectable by the LAT detector.

2.3. *GLAST sensitivity for GRB*

As final exercise, we compute the sensitivity of the LAT detector for GRB. We have simulated 10000 bursts with different inverse Compton components varing the parameter τ uniformly from 0 (pure synchrotron spectrum) to 10 (high inverse Compton component). Each burst has been normalized between 20 keV and 1 MeV to the BATSE fluence distribution. In this computation we consider the effective area flat above 100 MeV. For normal incidence photons we consider $A_{on-axis} = 0.84\ m^2$, and $A_{off-axis} = 0.05\ m^2$ (which corresponds to an incident direction of 67° degrees from the normal). The distribution of the the fluence in the BATSE energy range versus the number of generated photons above 100 MeV for 10000 simulated bursts is shown in left panel of Fig. 4. The two horizontal lines represent the burst detection thresholds for on-axis and off-axis incident directions: they represent the number of photons per square meter in order to have at least 5 photons detected by the LAT: these are the required number of events in order to recognized a spatial-temporal cluster needed to detect a transient signal[16]. The efficiencies in the GRBs detection are 89.4% for on-axis and 46.9% for off-axis, as depicted in the right panel of Fig.4. Considering that BATSE has detected 550 bursts per year over 4π and the field of view for on-axis and off-axis detection is equal to $2\pi \cdot 0.2$: we predict that the LAT will see from 30 to 50 bursts on-axis per year, (10 to 30 for off axis detection) depending on their spectral features. Left panel of Fig. 5 shows

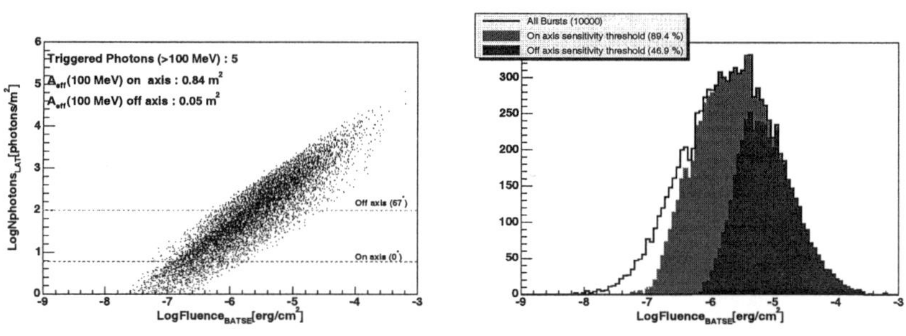

Figure 4. Left: number of generated photons per square meter above 100 MeV versus the fluence in the BATSE energy range. The two horizontal lines represent the burst detection thresholds for on-axis and off-axis observation. They represent the number of photons per square meter to generate in order to have at list 5 photons detected by the LAT. They are at $\log(5/0.84) = 0.77$ for on-axis detection, and and $\log(5/0.05) = 2$ for off-axis detection. The dots above the thresholds are detectable bursts. Right: detectable bursts ratio (filled histograms, for on-axis and off-axis) with respect to the total number of burst generated (empty histogram).

the fraction of GRBs detected as a function of the parameter τ, for on-axis detection and for off-axis detection.

Another interesting study that can be done using this approach, is to determine the distribution in redshift, and the highest observable redshift

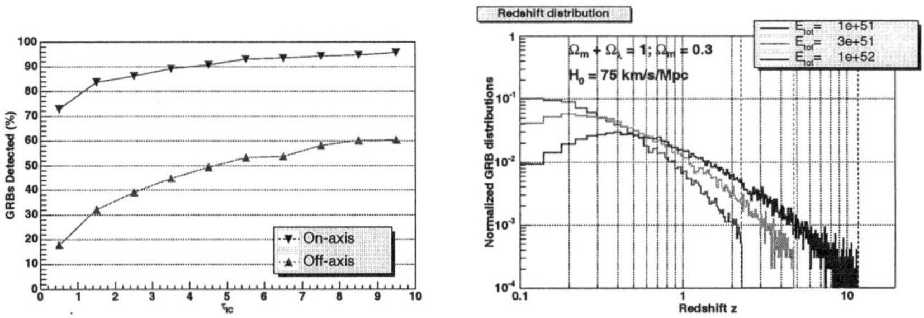

Figure 5. Fraction of detected GRBs as a function of the parameter τ. The situation of pure synchrotron spectrum yields a detection percentage of 73% for on-axis detection, and 18% for off-axis detection. For high inverse Compton component the on-axis detection approaches the 96% of the total generated bursts, while, for off-axis detection, the 61%. Simulated redshift distribution for different values of the intrinsic energy, and a given cosmology.

assuming GRB as standard candle[17,18,19]. Assuming a flat universe ($\Omega = \Omega_m + \Omega_\lambda = 1$, with $\Omega_m = 0.3$, and $H_0 = 0.71^{+0.04}_{-0.03}$ km/s/Mpc)[20,21] one can compute the redshift of a burst given the fluence F and assuming different values for the intrinsic energy E_{tot}:

$$z(F) = \frac{8\ c\ H_0 \sqrt{\pi\ E_{tot}\ F\ \Omega_m} + E_{tot} H_0^2 \Omega_m}{16 \pi c^2 F} \tag{1}$$

Right panel of Fig. 5 shows the distribution of redshift for three different values of the emitted energy. The intrinsic energy E_{tot} is assumed to be corrected by the beaming angle, so that even if the "isotropic equivalent" energy covers almost three decades of energy (10^{51} erg-10^{54} erg), the emitted energy is a narrow distribution centered at 1.3×10^{51}[18]. We can conclude that, depending on the Gamma-Ray Bursts energetic reservoir, GLAST will be able to scan the universe up to early epoch (z=10 for an intrinsic luminosity of 10^{52} ergs, z=5 for 10^{51} and z=2 for 3×10^{51} erg.

3. Conclusions

GLAST, the new generation gamma ray telescope, represents a big step forward in terms of technology and development, which will be translated in an enormous amount of new science, including discoveries. GRB are suitable sources for the GLAST mission: the LAT detector will observe the purely observed high energy range, adding informations on the acceleration of particles and on the radiative processes that allows production of gamma-rays up to GeV energies. In this article we simulated GRBs using a model that allows the flux normalization at BATSE energies and extrapolates the spectrum up to LAT energies, using the synchrotron model prescription. We shown that LAT is sensitive to the synchrotron cut-off: LAT will be able to reconstruct this spectral feature with statistic significance. We also simulate the inverse Compton component with a parametric, but accurate, spectral function. The presence of such component enhance the spectrum at LAT energy as well as determines a second bump in the $\nu F(\nu)$ spectrum. We finally investigate the sensitivity of the LAT to GRBs, computing the ratio of bursts which are independently detectable by the LAT instrument. LAT will see tens of bursts per year, depending on their spectral features. Gamma-Ray Bursts science with GLAST will definitely be one of the most exciting adventures, especially thanks to the Large Area Telescope.

References

1. R. W. Klebesadel, I. B. Strong, and R. A. Olson, *Ap. J. Lett.*, **182**:L85+, (1973).
2. E. Costa, *et al., Nature*, **387**:783–785, (1997).
3. B. L. Dingus, *Astrophys. & Space Sci.*, 231:187–190, (1995).
4. K. Hurley, *et al., Bulletin of the American Astronomical Society*, **26**:881–+, (1994).
5. K. Hurley, *et al., IAUCirc*, **5940**:1–+, (1994).
6. M. M. González, *et al., Nature*, **424** :749–751, (2003).
7. T. Piran.*Physics Reports*, **314**:575–667, (1999).
8. R. D. Blandford and C. F. McKee, *Physics of Fluids*, **19** :1130–1138, (1976).
9. Piran, T., Reviews of Modern Physics, **76**, 1143, (2005).
10. N. Omodei. Modeling, simulation and optimization of the detection of Gamma-Ray Burst with GLAST, *Ph.D. Thesis*, (2005).
11. P. Boinee, *et al.,* In *Science with the New Generation of High Energy Gamma-Ray Experiments : Between Astrophysics and Astroparticle Physics*, (2003).
12. W. S. Paciesas,*et al., Ap. J. Supp.*, **122**:465–495, (1999).
13. O. C. de Jager and A. K. Harding, *Ap. J.*, **396**:161–172, (1992).
14. O. C. de Jager, *et al., Ap. J.*, **457**:253–+, (1996).
15. R. Sari, T. Piran, and R. Narayan, *Ap. J. Lett.*, **497**:L17+, (1998).
16. J. Cohen-Tanugi *et al., astro-ph/***0301356**
17. D. A. Frail, *et al., Ap. J. Lett.*, **562**:L55–L58, (2001).
18. J. S. Bloom, D. A. Frail, and S. R. Kulkarni *Ap. J.*, **594**:674–683, (2003).
19. G. Barbiellini, *et al., Mon. Not. RAS*, **350** :L5–L8, (2004).
20. C. L. Bennett, *et al., Ap. J. Supp.*, **148**:1–27, (2003).
21. F. K. Hansen, A. Balbi, A. J. Banday, and K. M. Górski, *Mon. Not. RAS*, **354**:905–912, (2004).

OBSERVATIONS OF BLAZARS AND EGRET SOURCES WITH INTEGRAL

V. VITALE[(1)], M. DADINA[(1)], G. DI COCCO[(1)], L. FOSCHINI[(1)], F. GIANOTTI[(1)], L. MARASCHI[(2)], E. PIAN[(3)], F. SCHIAVONE[(1)], G. TAGLIAFERRI[(2)], A. TREVES[(4)]

(1) INAF / IASF, Sezione di Bologna, Italy - (2) INAF / OAB, Milano, Italy - (3) INAF /OAT, Trieste, Italy - (4) Università dell'Insubria, Como, Italy

The IBIS imager on board INTEGRAL provides a unique opportunity in searching for sources within large error circles or to monitor a large number of sources because of its large field of view (29x29 square degrees), associated with a fine point source location accuracy (few arcmin up to ~100 keV). Two programs are currently performed, by means of the regular scan of INTEGRAL pointings. The first, which is also part of the INTEGRAL Core Programme activity, is devoted to the search for possible counterparts of EGRET unidentified sources. The second one is the monitoring of a large sample of blazars coupled with the possibility to perform target-of-opportunity observations in case of significant outburst events. The two programs share a common strategy, both from the scientific and observational feasibility aspects. In this work, the activity status and the first results from the two programs are reported, and discussed in view of the GLAST mission perspectives.

1. EGRET Sources

More than half of the EGRET sources (E >100 MeV) are of unknown type. Their identification is one of the challenges in contemporary astrophysics. Blazars and pulsars are the most common candidate counterparts for EGRET unidentified sources. Among 271 sources contained in the 3rd EGRET catalog, 66 are blazars (plus 27 at lower statistical confidence) and 5 are pulsars. Nevertheless other populations are not excluded:- radiogalaxies such as Cen A, supernova remnants of shell type, isolated neutron stars, OB stars associations. All these candidates are "extreme" astrophysical systems, which can efficiently accelerate charged particles (E > 1012 eV). Thus they may emit broad-band non-thermal radiation. With a multi-wave-length (m.w.l.) study, it is possible: - to find candidate counterparts for the high energy gamma sources; - to reconstruct the broad-band emission spectra; - to investigate the temporal features of the sources. The search for counterparts of EGRET unid. sources with INTEGRAL has already given positive results (3EG J1736-2908 related to the AGN GRS 1734-292 [Di Cocco, et al. 2004 A&A 425,89] and 3EG J 1621+8204 related to the FR I radio galaxy N GC 6251 [Foschini et al. 2005 A&A 433,515]).

2. Blazars

Blazar type AGN are among the most violent phenomena in the universe. Blazars are very luminous broad-band emitters and show rapid variability. The gravitational compact energy source of AGNi may power relativistic jets of plasma. For blazars, such jets are seen at small angle of sight, and the observed emission is enhanced by relativistic effects. Shocks, propagating along the jets, might cause rapid variability. Jets structure and physics can be understood by means of the time-variability study. While outburst development and energy extraction (from the central energy source) can be investigated with time-resolved m.w.l. emission spectra. These are the motivations for monitoring blazars with IBIS. Furthermore, it might be possible to separate the contributions of synchrotron and inverse Compton to hard-X soft-gamma emission of blazars. The results obtained to date are the detections of S5 0716+714 [Pian et al. 2005 A&A 429, 427] and of 3C 454.3 [Foschini et al. 2005 ATel 497].

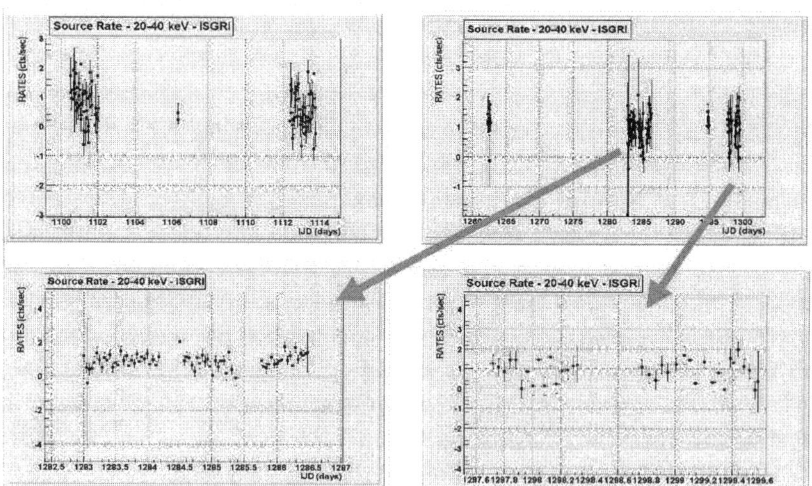

Figure 1. 3C273 LIGHT CURVE, (ISGRI counts/sec versus INTEGRAL Julian Date), the time binning is 2ksec.

3. Populations Study

We started a systematic study of the possible populations of EGRET sources. We first focused on the microquasar class of objects. The first tentative we made was to compare the expectations based on IC models for High and Low Mass micro-blazars (i.e. microquasar with jets aligned with our line of sight, both HMM and LMM) with INTEGRAL results. Since few are the cases of X-ray (20-60 keV) detections within the EGRET error contours, we compared the obtained X-ray upper limits with theoretical SED calculated for both HMM and LMM. For this purpose, we analyzed the Galactic Center (GC) region data (maximum used exposure of 1.1 Ms). In this field, 44 Unidentified EGRET Sources are present.

We calculated the flux values needed to have a 3σ (Ubertini et al. 2004) detection with respect to the different exposures for the EGRET sources in the field, in order to get upper limits. Since mosaicing tecnique may introduce some systematics, we increased these values by a factor of 2.36 (Bird

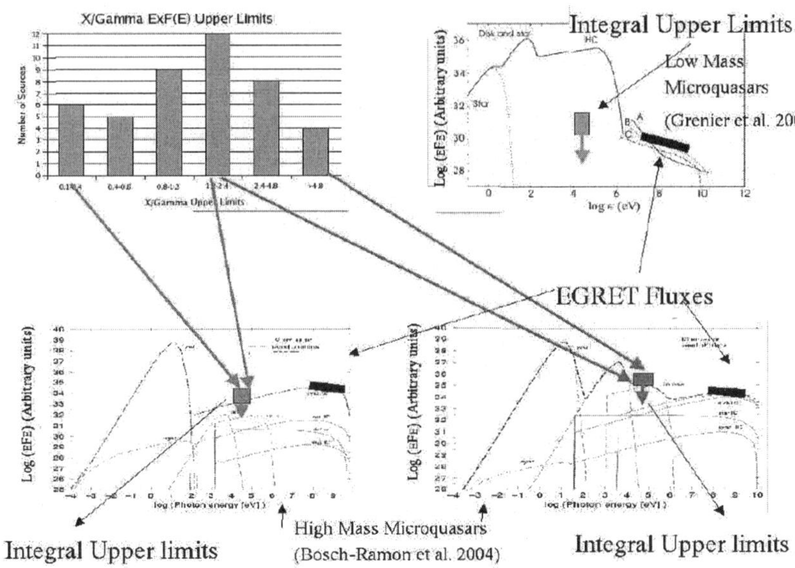

Figure 2. COMPARISON OF X/Gamma E*F(E) UPPER LIMITS WITH CALCULATED SED FOR HMM AND LMM. On the top-left panel are the ratios of upper limits of ExF(E) in the IBIS energy band and the E*F(E) in the EGRET energy band. In the other three panels, the IBIS upper-limits are compared with SED calculated for HMM and LMM in Grenier et al. 2005 and Bosch-Ramon et al. 2004.

et al. 2004, Galactic plane survey statistical study), in order to estimate a "realible" 95% upper limit. Then we calculated the upper limits on the X/Gamma ratios. These values have been compared with those expected from theoretical model. Our data are in line with what expected for HMM. From our results, some cases are also in agreement with what expected for objects not completely dominated by Jet emission. On the contrary, the data do not seem to agree with the models for LMM. Present results seem to suggest that LMM could be candidate for unidetified EGRET sources only if the geometry of the observer-source system is extreme (observer line of sight within 1deg from the jet direction).

4. Methods of Study and Perspectives

Gamma rays sources are observed, with IBIS, in the energy band between 15 keV and 10MeV (Ubertini et al., 2003 A&A 411, 131). IBIS provides: - an angular resolution $\delta\Theta = 5'$ between 15 keV and 1 MeV (IBIS/ISGRI, Leburn et al., 2003 A&A 411, 141), and $\delta\Theta = 10'$ between 175 keV and 10 MeV (IBIS/PICsIT, Di Cocco et al., 2003 A&A 411, 149); - a large field of view (29*29 deg^2, 19*19 deg^2 at half response), which allows one to study many sources with a single pointing. INTEGRAL cumulates exposures of 10^7 secs/year by means of source pointings and scan programs (Galactic Plane Scan, Galactic Centre Deep Exposure). Outbursting sources, such as blazars, can be detected within a single pointing (2 ksec). Steady weak sources can, instead, be detected, down to few mCrab, with exposures of 10^6sec, in the energy band 20-100 keV. The main steps of the data analysis are: 1) the building up of sky maps (in 3 bands: 20-40, 40-100, 200-400 keV), for each single pointing; 2a) the study of source light curves, for the monitored blazars; 2b) the search for signals within the error boxes of EGRET unidentified sources; 3) the study of weak sources, by cumulating large exposures of the source under study (stacking of the single relevant pointings); 4) further m.w.l. study of the detected sources. Approximately 6.5*10^6 seconds (from revolution 23 to 148) of data have been processed, and are ready for further investigations. The m.w.l. studies will benefit from the Gamma Ray Large Area Space Telescope (GLAST). This satellite will cover the 20 MeV to 50 GeV energy range, and it is scheduled for launch in 2006. With GLAST data one will have: a) a better location of gamma sources; b) detection of weaker objects; c) more accurate spectra; d) the opportunity of simultaneous observations from the hard-X (as example with INTEGRAL) to the high energy gamma rays.

A THIRD LEVEL TRIGGER PROGRAMMABLE ON FPGA FOR THE GAMMA/HADRON SEPARATION IN A CHERENKOV TELESCOPE USING PSEUDO-ZERNIKE MOMENTS AND THE SVM CLASSIFIER

MARCO FRAILIS, ORIANA MANSUTTI,
PRAVEEN BOINEE, GIUSEPPE CABRAS, ALESSANDRO DE ANGELIS,
BARBARA DE LOTTO, ALBERTO FORTI

Dipartimento di Fisica, Università di Udine
& INFN Sezione di Trieste - Gruppo collegato di Udine,
Via delle Scienze 208 – 33100 Udine, Italy
E-mail: deangelis@fisica.uniud.it, frailis@fisica.uniud.it, mansutti@fisica.uniud.it

MAURO DELL'ORSO

Università di Pisa
Lungarno Pacinotti, 43 – 56126 Pisa, Italy

RICCARDO PAOLETTI, ANGELO SCRIBANO, NICOLA TURINI

Università di Siena
via Banchi di Sotto, 55 – 53100 Siena, Italy

MOSÈ MARIOTTI, LUIGI PERUZZO, ANTONIO SAGGION

Università di Padova
via 8 Febbraio, 2 – 35122 Padova, Italy

We studied the application of the Pseudo-Zernike features as image parameters (instead of the Hillas parameters) for the discrimination between the images produced by atmospheric electromagnetic showers caused by gamma-rays and the ones produced by atmospheric electromagnetic showers caused by hadrons in the MAGIC Experiment. We used a Support Vector Machine as classification algorithm with the computed Pseudo-Zernike features as classification parameters. We implemented on a FPGA board a kernel function of the SVM and the Pseudo-Zernike features to build a third level trigger for the gamma-hadron separation task of the MAGIC Experiment.

1. Introduction

Gamma-ray detectors experiments have the need of separating signals produced by photons from signals produced by hadrons.

In the case of the MAGIC telescope, the need is to discriminate between the images produced by atmospheric electromagnetic showers caused by gamma-rays and the ones produced by atmospheric electromagnetic showers caused by hadrons.

To contrubute to this discrimination, we studied some approaches applicable to the gamma-hadron separation of the MAGIC telescope experiment.

We set up our work as a two-class image classification problem. The novelties we propose lie in two phases of the consutruction of a classifier for the MAGIC images: in the choice of the classification parameters and in the choice of the classification algorithm.

We used the results of these classification problem to build a third level trigger for the gamma-hadron separation task of the MAGIC Experiment by implementing a kernel function of the classificator and the pseudo-Zernike featrues on a FPGA board, which can be used to discriminate the images coming from the MAGIC telescope.

2. Classification Parameters

The image parameters usually used for image classification and analysis in the MAGIC telescope collaboration are the Hillas parameters [1,2].

In our work we considered the application of the pseudo-Zernike features [3] as image parameters for the MAGIC telescope images.

The *Zernike moments* [4] are an infinite orthogonal basis of polynomials for the representation of image functions. The value of the coefficient of a basis element is referred to as *Zernike feature* of the image.

A first characteristic of this basis that makes it of interest for the MAGIC telescope images is the rotation invariance of the Zernike moments, that makes the representation of the image independent from a rotation of the telescope along its symmetry axis.

Two other characteristics of Zernike moments that are important for the MAGIC telescope images are that they are not scale and translation invariant (though they can be modified to get these invariances):

- avoiding the scale invariance makes these parameters sensitive to the magnitude of the image, and this means that the further classification will take into account the energy of the originating particle (and the classifier automatically behaves differently for low energy and high energy images);

- avoiding the translation invariance makes these parameters sensitive to the Hillas alpha parameter of the image (the transformation that leads to translation invariance would make all alphas equal to zero).

The orthogonality property of Zernike moments enables one to select the required number of features to get a good enough representation of the image, since orthogonality:

- makes the image reconstruction from its features computationally simple;
- enables one to evaluate the image representation ability (contribution to the reconstruction process) of each order moment.

Unfortunately, high order Zernike moments are very sensitive to noise [5]. In order to solve this problem and keep all the useful characteristics of Zernike moments, we have considered the pseudo-Zernike moments.

Like the Zernike moments, also the *pseudo-Zernike moments* [3,6] are an infinite orthogonal basis of polynomials for the representation of image functions (where the value of the coefficient of a basis element is referred to as *pseudo-Zernike feature* of the image), and are not scale and tanslation invariant.

Pseudo Zernike moments are more robust to image noise, since the number of pseudo-Zernike polinomials of a fixed maximum order n is $(n+1)^2$, while the Zernike moments are in number of $\frac{1}{2}(n+1)(n+2)$ [7].

3. Classification Algorithm and Training Strategy

Both with the choice of the Hillas parameters and the pseudo-Zernike features as classification parameters, a classification algorithm has to be chosen (for which the selected parameters are the input).

The classification algorithm we used is based on the usage of a Support Vector Machine (SVM) as classification technique [8], since they are currently under very active research within the fields of machine and statistical learning and they had not been conclusively tested for our classification task [9]. The SVM classification method is systematic, reproducible, and properly grounded by the statistical learning theory.

The SVM algorithm solves an optimization problem on a set of training data to determine the parameters of a function (the *decision function*) to be evaluated on the data that have to be classified. The data have to be described through several features (in our case the Hillas parameters or the pseudo-Zernike features) and the evaluation of the decision function must give a target value that represents the result of the classification task (in our case, a value telling if the the image was produced by a primaty photon or hadron, that can be represented by a value in $\{1, -1\}$) [10,11].

In its simplest form, an SVM is able to perform a binary classification finding the "best" separating hyperplane between two linearly separable classes. There are infinite hyperplanes properly separating the data. So, the SVM finds this hyperplane maximizing the distance, or *margin*, between the *support hyperplanes* for each class. A hyperplane supports a class if all points in that class are on one side of that plane. This problem is formulated as a quadratic programming problem (QP) and can be solved by effective robust algorithms. If the data is not linearly separable, slack variables are introduced into the QP problem to accept outliers.

In order to learn non-linear relations, SVM implicitly applies a fixed non-linear mapping of the data to a enough high (maybe infinite) dimensional feature space in which the classification through the hyperplanes can be used [10]. The implicit mapping is performed by using the so called *kernel functions*, that one can choose according to which function makes the SVM perform better [11]. In the leterature, several kernel functions have been proposed [10]. Among them, the Gaussian radial basis function kernel is tipically used for classification tasks. It is defined by:

$$k(x, y) = e^{-\gamma \|x-z\|^2}$$

where x and z are input data and γ is a parameter specified by the user regulating the width of the Gaussian kernel.

4. Training the SVM algorithm

The SVM algorithm with the Gaussian kernel requires two parameters: the parameter $C > 0$ is a regularization constant determining a trade-off between the empirical error (number of wrongly classified inputs) and the *complexity* of the found solution; the γ parameter of the Gaussian kernel affects the complexity of the decision boundary.

To select the best C and γ parameters for the training step, we used a cross validation via grid search. With this method, the training set is first separated into m folds. Sequentially a fold is considered as the validation set and the rest are for training. So an adaptive grid-search is performed on C and γ, trying exponentially growing sequences for the two parameters.

5. Case study: gamma/hadron separation in MAGIC

To test both the pseudo-Zernike features and the SVM classification algorithm for the gamma/hadron separation problem in a Cherenkov telescope we used the software infrastructure built for the MAGIC experiment.

First, a dataset of simulated gamma-ray photons has been produced, generating calibrated Cherenkov images for each event, with different pointing positions. Then, real OFF data (i.e. data taken from a sky region not too far away from the Crab source) gathered by the MAGIC telescope has been used to represent the background (hadrons, electrons, muons, diffuse photons).

From these dataset we extracted a training set of 12228 gammas and 12306 hadrons and a test set of 6109 gammas and 6183 hadrons (using a random sampling). We considered events taken with pointing position comprised between 0 and 12 degrees.

We modified the MAGIC software pipeline to extract the Zernike features from each Cherenkov image but using the standard image cleaning method currently adopted in MAGIC to select only the pixels above the pedestal signal.

After extracting the pseudo-Zernike features from each image (we use an order $n = 7$, corresponding to 36 features), we normalize the training and test set with respect to the mean and standard deviation of each feature (calculated on the training set). This is performed to avoid features in greater numeric ranges dominate those in smaller numeric ranges. With this method, each feature varies approximately in the range $[-1, 1]$.

After applying the cross validation via grid search (see Figure 1) on a small randomly sampled subset of the training set (5% of the entire set), we have obtained the best rates for $C = 28526.2$ and $\gamma = 1.07$. Finally, we trained the SVM classifier with the aforementioned parameters obtaining the following results on the test set:

	Total	Recognized	Ratio
Gammas	6109	5271	86.3%
Hadrons	6183	4259	68.9%

The overall accuracy obtained is 77.5%.

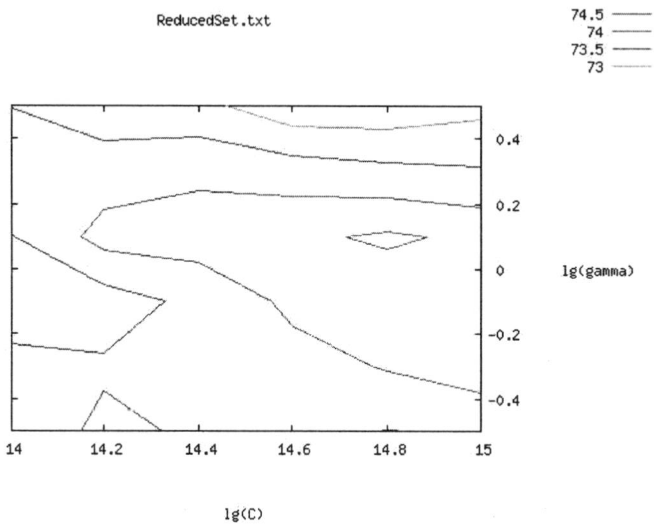

Figure 1. Cross validation with grid search.

6. FPGA implementation

A Xilinx Spartan 3 FPGA board was chosen to implement both the feature extractor (computing the pseudo-Zernike moments) and the SVM decision function while the training algorithm is meant to be performed off-line via software. Alternative FPGA-based implementations of Zernike moments and Support Vector Machines can be found in [14,15].

As programming language we chose a commercial product, ImpulseC, a C to RTL/HDL compiler (see Figure 2). With this approach we can reason in terms of algorithms, not of the hardware logic, requiring less efforts to convert the original C program to the FPGA device [16].

Currently, ImpulseC does not support all the features of the C standard language for the hardware translation. In particular, it only supports integer arithmetic and fixed point basic arithmetic. No global variables or function calls can be used within the main function to be translated in VHDL.

Therefore, we had to implement from scratch the fixed point square root (to calculate zernike features) and the exponential function (used by the SVM decision function). To avoid function calls, we implemented them as C macros. To improve the efficiency of the pseudo-Zernike features

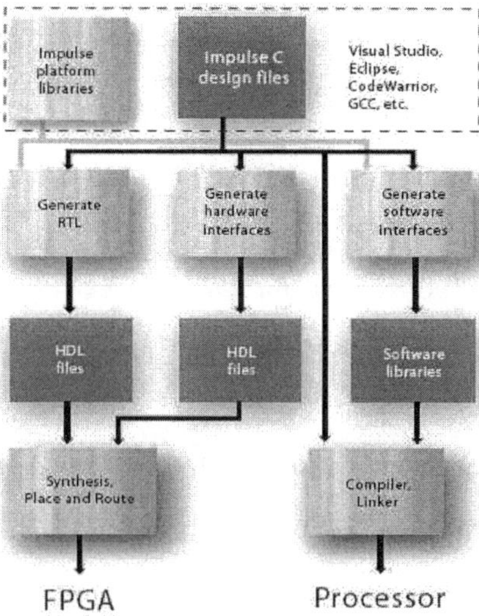

Figure 2. C to HDL conversion with ImpulseC.

computation and reduce the approximation errors (due to the fixed point arithmetic), the pseudo-Zernike radial polynomials, which depend only on the pixel coordinates of the telescope, are calculated only once via software and stored as fixed point values in the FPGA.

References

1. A. M. Hillas, Proc. 19th ICRC, La Jolla 3 (1985) 445
2. W. Wittek, MAGIC-TDAS 02-03 (2002) 020131
3. A. Bathia, E. Wolf, Proceedings of Camb. Phil. Soc., **50** (1954) 40
4. A. Khotanzad, Y. H. Hong, IEEE Transactions on Pattern Anslysis and Machine Intelligence **12**, **5** (1990) 489
5. S. X. Liao, M. Pawlak, IEEE Transactions on Pattern Anslysis and Machine Intelligence **20**, **12** (1998) 1358
6. R. Mazza, *Un sistema per il riconoscimento di comandi gestuali*, tesi di laurea (laurea degree thesis), Universitï¿½di Pisa (1997)
7. C.-W. Chong, R. Mukundan, P. Raveendran, Proc. of Intnl. Conf. on Image and Vision Computing - IVCNZ'01, Newzeland (2001) 237
8. C. J. C. Burges, *A Tutorial on Support Vector Machines for Pattern Recognition*, in *Data Mining and Knowledge Discovery* **2**, Kluwer (1998) 121
9. R. Bock *et al.*, Nuclear instruments and Methods in Physics Research A **516** (2004) 511
10. C.-W. Hsu, C.-C. Chang, C.-J. Lin, *A Pratical Guide to Support Vector*

Classification,
`http://www.csie.ntu.edu.tw/~cjlin/papers/guide/guide.pdf`
11. K. P. Bennett, C- Campbell, SIGKDD Explorations **2**, **2** (2000) 1
12. B. Schölkopf, A. J. Smola, R. C. Williamson, P. L. Bartlett, Neural Computation **12**, (2000) 1207
13. C.-C. Chang, C.-J. Lin, *LIBSVM: a Library for Support Vector Machines*, `http://www.csie.ntu.edu.tw/~cjlin/papers/libsvm.pdf`
14. L. Kotoulas, I. Andreadis, *Real-time computation of Zernike moments*, IEEE Trans. on Circuits and Systems for Video Technology, Volume: 15, Issue: 6, pages 801- 809 (2005)
15. D. Anguita, A. Boni, A. Zorat, *Mapping LSSVM on Digital Hardware*, Proc. of the IEEE Int. Joint Conf. on Neural Networks, IJCNN 2004, Budapest, Hungary (2004)
16. D. Pellerin, S. Thibault, *Practical FPGA Programming in C*, Prentice Hall PTR (2005)

PULSARSPECTRUM: SIMULATING GAMMA-RAY PULSARS FOR THE GLAST MISSION

MASSIMILIANO RAZZANO, LUCA LATRONICO, NICOLA OMODEI, GLORIA SPANDRE

Istituto Nazionale di Fisica Nucleare, sezione di Pisa,
Largo B.Pontecorvo 3, 56100 Pisa
E-mail: massimiliano.razzano@pi.infn.it

We present here an overview of *PulsarSpectrum*, a program that simulates the gamma ray emission from pulsars. This simulator reproduces not only the basic features of the observed gamma ray pulsars, but it can also simulate more detailed effects related to pulsar timing. It is a very useful tool to understand the GLAST capabilities in the pulsar science.

1. Introduction

Pulsars are among the most exciting gamma ray sources in the Universe and can serve as unique sites for the study of emission processes in extreme physical environments. The Gamma ray Large Area Space Telescope (GLAST) will increase dramatically our knowledge of gamma ray pulsars physics. In particular the Large Area Telescope (LAT), the main GLAST instrument, will provide more detailed observations of the known gamma ray pulsars and potentially will discover many new pulsars that emit gamma rays. To better understand the capabilities of GLAST for pulsar science we developed *PulsarSpectrum*, a program that simulates gamma ray emission from pulsars. This simulator can be easily interfaced with the Monte Carlo software that simulates the response of the LAT.

1.1. *Gamma ray pulsars in the GLAST era*

Pulsars have been associated with high energy gamma ray astronomy since the first experiment capable to resolve point sources. The *Small Astronomy Satellite*(SAS-2), launched in 1972, identified three intense sources in the sky, that in the following turned out to be the Vela, Crab and Geminga pulsars. A signal modulated at the radio period was found for Vela [1] and

Crab [2], but for Geminga no radio counterpart was found and this source remained quite mysterious until it was identified as an X-ray and gamma-ray pulsar. The *Energetic Gamma Ray Experiment* (EGRET) aboard the *Compton Gamma Ray Observatory* (CGRO) detected three more pulsars, B1706-44 [3], B1055-52 [5] and later B1951+32 [9]. Another one, B1509-58 was observed at energies between 60 keV and 2 MeV by the three CGRO experiments, the *Burst and Transient Source Experiment* (BATSE)[7], the *Oriented Scintillation Spectrometer Experiment* (OSSE), [6] and the *Imaging Compton Telescope* (COMPTEL)[4], but it was not seen by EGRET, because the spectrum shows a cutoff at energies lower than the instrument's threshold. Recently the millisecond pulsar PSR 0218+4232 was detected by analyzing the EGRET data [10], increasing to eight the number of detected gamma ray pulsars, seven of them are radio-loud pulsars and only Geminga seems to be radio quiet.

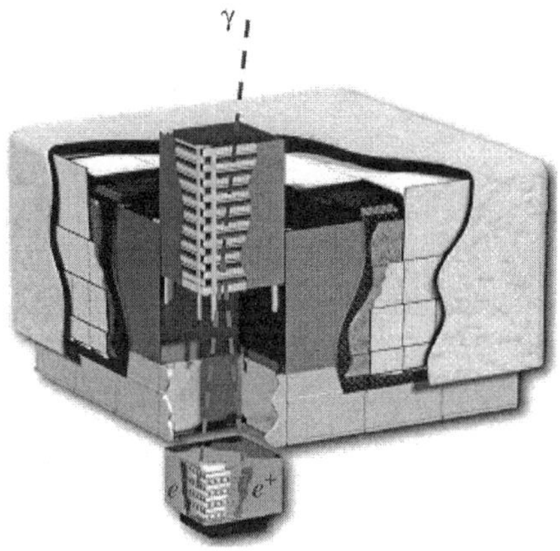

Figure 1. The GLAST Large Area Space Telescope. An incident gamma ray converts into a e^-e^+ pair

The GLAST Large Area Telescope (LAT) (fig 1), is a pair conversion telescope based on the most advanced high energy detectors. It consists in a precision silicon strip tracker, a hodoscopic calorimeter and a segmented anticoincidence shield for particle background rejection. The LAT high sensitivity (2 10^{-9} ph/cm^2/s) and effective area (>8000 cm^2) will permit the discovery of a lot of new pulsars: the estimates range between tens to hundreds depending upon the model adopted. Moreover the low dead time of the detector (20 μs) will allow the detailed reconstruction of pulsar lightcurves. One of the most exciting possibilities of the LAT will be the coverage of the energy window from 30 GeV up to 300 GeV, a still unexplored range. At these energies the theoretical models make different previsions on the high energy spectral cutoff and the spectral coverage of LAT will be of primary importance for constraining and discriminating among the models. In order to study the LAT response to specific gamma ray sources, various simulation packages have been developed. Here we present *PulsarSpectrum*, that can simulate the observed pulsars and create new fake pulsars for the LAT threshold detection identification.

2. The *PulsarSpectrum* simulator

2.1. *Overview of the simulator*

The basic idea behind *PulsarSpectrum* is to construct a 2-dimensional histogram representing the differential flux vs. energy and pulsar phase. This histogram contains all the basic informations about lightcurve and spectrum. How it is built depends upon the model we want to use: a phenomenological model, based only on observations, or a physical one, based on a specific theoretical model. At present only a phenomenological model has been implemented because it is more flexible and completely independent from the chosen theoretical scenario. The input parameters of the simulator can be divided in two categories:

- *Observational parameters*, (i.e the flux or the ephemerides);
- *Model-dependent parameters*, (i.e. the spectral index);

These parameters are placed in two specific data files used by both *PulsarSpectrum* and the LAT simulation tools. *PulsarSpectrum* creates the lightcurve and the spectrum from these parameters and combines them to obtain a two-dimensional matrix that represents the flux in ph/m^2/s/keV. The photons are then extracted such that the interval between two subsequent photons is determined by the flux integrated over the energy range of interest.

2.2. *The phenomenological model*

The currently implemented phenomenological model allows the user to generate pulsar lightcurves in a general way using a single or double Lorenzian peak profile whose shape is determined from random generated numbers. The lightcurve can be generated alternatively from a user-provided profile. This is useful for simulating the already observed gamma ray pulsars.

The spectral shape is assumed to be a power law with exponential cutoff (according to [8]), as in the observed gamma-ray pulsars and can then be modeled as:

$$\frac{dN}{dE} = K(\frac{E}{E_n})^a exp(\frac{E}{E_0})^{-b} \tag{1}$$

The normalisation constant K is determined by the photon flux above 100 MeV and the other four parameters can be varied; the values for the EGRET pulsars are obtained from real data by fitting procedures (e.g. [8]).

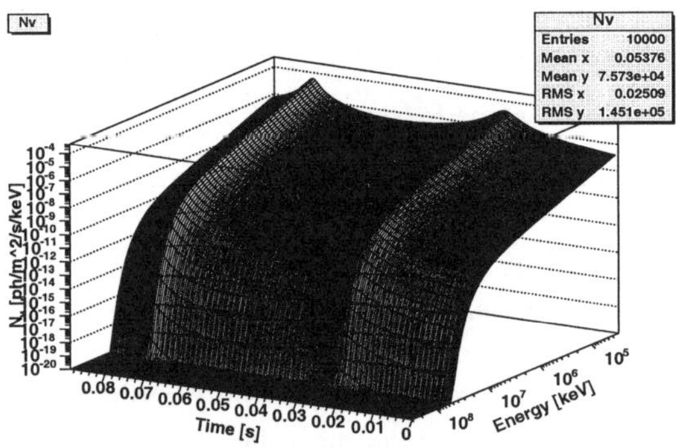

Figure 2. An example of 2-dimensional histogram created by PulsarSpectrum.

2.3. *Timing issues*

Once the differential flux histogram is created the time interval between two subsequent photons is computed according to the flux. If the previous photon came at time t_0 the next photon will appear at $\tilde{t}$ such that:

$$A_{eff} \int_{t_0}^{\tilde{t}} \int_{E_1}^{E_2} \frac{dN}{dEdAdt} dEdAdt = 1 \tag{2}$$

The interval between two photons is computed assuming that the pulsar period does not change with time and the photons arrival times are computed into a reference system fixed relative to stars. This is not the "real world". Pulsar timing is affected by more complicate effects, as (1)- The motion of the spacecraft through the Solar System and the relativistic effects due to gravitational well of the Sun (see 2.3.1); (2)- Period changes with time (see 2.3.2). For pulsars in binary systems an additional modulation to the orbital period should be taken into account. For a precise pulsar simulator intent to produce a realistic list of photon arrival times we need to include all these effects (to transform to the observational frame). All these procedures are now implemented in the code and only the binary demodulation is not yet implemented. The real arrival time of a photon from a pulsar must be first barycentered and then phase assigned.

2.3.1. *Barycentric effects*

The first step to analyze pulsar data is the conversion from the arrival times at the spacecraft, usually expressed in Terrestrial Time TT or TAI, to the arrival times at the Solar System barycenter, expressed in Barycentric Dynamical Time TDB. Taking into account both the motion of spacecraft through space and the general relativistic effects due to the gravitational field of the Sun (i.e. Shapiro delay), the simulator computes the opposite of the barycentric correction by considering the position of the Earth and of the spacecraft in the Solar System, and the position of the Sun. The accuracy for the computation of these "de-corrections" is hard-coded in the program.

2.3.2. *Period change and ephemerides*

The rotational energy of a radio pulsar decreases with time and hence the period increases with time. For gamma ray pulsar science the radio ephemerides are fundamental for assigning the correct phase to each photon. If we know the frequency $f(t_0)$ and its derivatives $\dot{f}(t_0)$ and $\ddot{f}(t_0)$ at a certain time t_0, known as *epoch*, the phase is then:

$$\phi(t) = int[f(t_0)(t - t_0) + \frac{1}{2}\dot{f}(t_0)(t - t_0)^2 + \frac{1}{6}\ddot{f}(t_0)(t - t_0)^3]. \quad (3)$$

The interval between two photons must be "de-corrected" for this effect. In the parameters file the user can specify a set of ephemerides with the relative epoch of validity expressed in Modified Julian Date. The simulator then computes the opportune arrival time such that, after applying the barycentric corrections and the equation 3, the correct phase is obtained.

3. Conclusions

PulsarSpectrum has been successfully used by the GLAST collaboration for testing the functionality of the LAT Science Analysis Tools, a set of analysis programs specifically designed to analyse the LAT data after launch. Periodically these tools are evaluated through specific Checkout phases. In

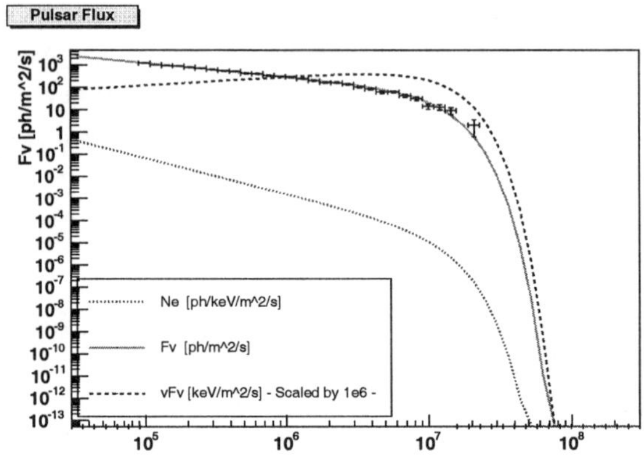

Figure 3. An example of the spectrum of a simulated pulsar created by PulsarSpectrum. The point represents the extracted photons of energy above 100 MeV.

absence of real data there is strong need of detailed simulated data, most of them provided by *PulsarSpectrum*. *PulsarSpectrum* is the most probable candidate to be used as pulsar simulator in the GLAST Data Challenge 2, an analysis phase where some months of LAT data are simulated. A new interesting opportunity is now raising from the creation of the LAT Science Groups, to study GLAST science opportunities on specific topics. Our simulator has all the characteristics to fit well the requirements of the LAT Pulsar Science Group.

References

1. D.A. Kniffen et al. *Nature* **251**, 397 (1974).
2. D.J. Thompson et al. *ApJ* **200**, L79 (1975).
3. D.J. Thompson et al.*Nature* **359**, 615 (1992).
4. K. Bennet at al.*Proc 23d Int. Cosmic Ray Conf (Calgary)* **1**, 172 (1993).
5. J.M. Fierro et al.*ApJ* **413**, L27 (1993).
6. M.P. Ulmer et al.*ApJ* **417**, 738 (1993).
7. R.B. Wilson et al.*Isolated Pulsars, ed. K.A. Van Riper,R.Epstein, and C.Ho (Cambridge:Cambridge Univ. Press)* 257 (1993).
8. H.I. Nel and O.C. De Jager *Astrophysics and Space Science* **230**, 209 (1995).
9. P.V. Ramanamurthy et al. *ApJ* **447**, L109 (1995).
10. L.Kuiper *Pulsar Astronomy, 2000 and Beyond, IAU Colloquium 177, Astronomical Society of Pacific, San Francisco* , 355 (2000).
11. X. Yan, *Phys. Rev.* **E61**, 4745 (2000).

INDEX OF AUTHORS